Transfusion Medicine
for Pathologists

Transfusion Medicine for Pathologists
A Comprehensive Review for Board Preparation, Certification, and Clinical Practice

Brian Castillo

Pathology and Laboratory Medicine, McGovern Medicine School, University of Texas
Health Science Center at Houston, Houston, TX, United States

Amitava Dasgupta

Pathology and Laboratory Medicine, McGovern Medicine School, University of Texas
Health Science Center at Houston, Houston, TX, United States

Kimberly Klein

Pathology and Laboratory Medicine, McGovern Medicine School, University of Texas
Health Science Center at Houston, Houston, TX, United States

Hlaing Tint

Pathology and Laboratory Medicine, McGovern Medicine School, University of Texas
Health Science Center at Houston, Houston, TX, United States

Amer Wahed

Pathology and Laboratory Medicine, McGovern Medicine School, University of Texas
Health Science Center at Houston, Houston, TX, United States

ELSEVIER

Elsevier
Radarweg 29, PO Box 211, 1000 AE Amsterdam, Netherlands
The Boulevard, Langford Lane, Kidlington, Oxford OX5 1GB, United Kingdom
50 Hampshire Street, 5th Floor, Cambridge MA 02139, United States

Notices
Knowledge and best practice in this field are constantly changing. As new research and experience broaden
our understanding, changes in research methods, professional practices, or medical treatment may become
necessary.

Practitioners and researchers must always rely on their own experience and knowledge in evaluating and
using any information, methods, compounds, or experiments described herein. In using such information
or methods they should be mindful of their own safety and the safety of others, including parties for whom
they have a professional responsibility.

To the fullest extent of the law, neither the Publisher nor the authors, contributors, or editors, assume any
liability for any injury and/or damage to persons or property as a matter of products liability, negligence
or otherwise, or from any use or operation of any methods, products, instructions, or ideas contained in
the material herein.

Library of Congress Cataloging-in-Publication Data
A catalog record for this book is available from the Library of Congress

British Library Cataloguing-in-Publication Data
A catalogue record for this book is available from the British Library

ISBN: 978-0-12-814313-1

For information on all Elsevier publications visit our website at
https://www.elsevier.com/books-and-journals

Working together
to grow libraries in
developing countries

www.elsevier.com • www.bookaid.org

Publisher: Mica Haley
Acquisition Editor: Tari K. Broderick
Editorial Project Manager: Megan Ashdown
Production Project Manager: Swapna Srinivasan
Designer: Miles Hitchen

Typeset by Thomson Digital

Contents

Preface

Preparing for examination conducted by American Board of Pathology is an arduous task as residents must have mastery in all subspecialities of clinical pathology including clinical chemistry, point-of-care testing, hematology, coagulation, microbiology, blood banking. The first book to help pathology residents to prepare for the board, *Clinical Chemistry, Immunology and Laboratory Quality Control: A Comprehensive Review for Board Preparation, Certification and Clinical Practice*, was published by Elsevier in January 2014. After receiving favorable reviews we wrote the second book in the series *Hematology and Coagulation: A Comprehensive Review for Board Preparation, Certification and Clinical Practice*, which was also published by Elsevier in February 2015. That book was also well received by pathology residents and others which inspired us to write the third book in the series *Microbiology and Molecular Diagnostics in Pathology: A Comprehensive Review for Board Preparation, Certification and Clinical Practice*, which was published by Elsevier in June 2017. Now we are completing this series with this fourth book *Transfusion Medicine for Pathologists: A Comprehensive Review for Board Preparation, Certification, and Clinical Practice*.

This book is designed to help residents to quickly review all important topics related to transfusion medicine before taking the pathology board examination. This book will also help first year residents to understand fundamental basis of transfusion medicine related to practice of clinical pathology and then to read text books on blood banking for broadening their knowledge base. Therefore, this book is not an alternative to any standard text book related to blood banking or transfusion medicine but rather a compendium to help the resident understand basis of transfusion medicine practice during their initial training phase and then to prepare for the pathology board examination.

This book has 12 chapters and provides a comprehensive review of all important aspects of transfusion medicine practice. We have added a section, denoted as "key points" at the end of each chapter. We hope that this section will be a good resource for reviewing information, when time at hand is somewhat limited just before taking the board examination. The first chapter in this book discusses donor selection criteria, acceptance and refusal of donors, as well as donor testing for infectious disease. Chapter 2 addresses various blood components derived from blood and proper utilization of such products. Chapter 3 discusses major transfusion reactions and appropriate further testing and management of such patients. Chapter 4 is devoted to various testings performed in blood bank service. Chapter 5 discusses various red cell antigens and antibodies while Chapter 6 reviews the indications as well as principle of apheresis, physiology, and side effects of common apheresis procedures such as therapeutic plasma exchange, red cell exchange, and photopheresis. Chapter 7 discusses appropriate use of blood products. Chapter 8 reviews alternative methods to blood product administration for management of bleeding patients. Chapter 9 discusses quality control and quality assurances related to operation of blood bank service as well as regulatory issues in transfusion medicine. Chapter 10 discusses situations such as neonatal alloimmune thrombocytopenia, intra uterine transfusions, massive transfusions, and hemolytic disease of fetus and newborn. Chapter 11 covers various pharmacological agents used in transfusion medicine while Chapter 12 is devoted to issue of error detection and correction in blood bank service.

    We would like to thank our Department Chair, Dr. Robert Hunter for encouraging us to write the fourth book in the series and his support during our long process of preparing the manuscript. If pathology residents and other readers find this book useful, our efforts will be duly rewarded.

Brian Castillo
Amitava Dasgupta
Kimberly Klein
Hlaing Tint
Amer Wahed

Donor selection and testing

INTRODUCTION

The major purpose of donor selection is to provide safest possible blood products to the recipients as well as to protect the donor if there is any risk involved in donating blood or donor has a health condition which is not suitable for donating blood. The blood center aims to protect the recipient from infective organisms, exposure from potentially teratogenic agents, and theoretical risk of transmission of illnesses such as malignancy.

Over the last 20–25 years, considerable improvement in providing safe blood to recipients have taken place in the United States as well as many other countries [1]. As a result, posttransfusion infections are steadily declining worldwide. However, risk-free supply of blood products has not been achieved yet. A donor can donate:

- One unit of whole blood
- Two units of red blood cells (RBCs)
- Platelets apheresis
- Plasma apheresis
- Granulocytes

The donated units are properly used for any suitable recipient. In addition, the donation may be autologous or directed. In autologous donation, a donor donates blood in advance for donor's own use. These units are not subject to the same testing criteria as other donations and thus, if not used will be discarded. Directed donations are those donations made for a specific individual. Such blood products are subjected all testing and if not used may be made available for other patients.

STEPS OF DONOR SELECTION

There are four major steps of donor selection:

1. Donor screening by means of a medical history interview
2. Reading of educational materials
3. Undergoing a brief physical examination
4. Laboratory testing

Transfusion Medicine for Pathologists. http://dx.doi.org/10.1016/B978-0-12-814313-1.00001-0

MEDICAL HISTORY INTERVIEW

A full-length donor history questionnaire (DHQ) developed by the American Association of Blood Bank (AABB) and approved by the Food and Drug Administration (FDA) is used by many blood centers for screening donors. This questionnaire may be administered orally or can be self-administered. Follow-up questions will be asked by donor center staff as and when necessary. Highlights of various questions in DHQ are listed in Table 1.1. The FDA has also approved an abbreviated donor history questionnaire (aDHQ) for qualified frequent donors. This is because some questions in the uniform donor questionnaire (UDHQ) refers to events in the remote past and recent changes in medical, behavior, and travel information will be identified by the aDHQ. Highlights of various questions in aDHQ are listed in Table 1.2. Both questionnaires are readily available from the AABB website. Some centers have a computer-assisted self-interviewing process that utilizes audio, visual, and touch-screen components. The potential donor should be provided adequate privacy when self-answering the questions.

Donor demographics are required which will be used to identify the donor for traceability and track ability. Donor demographics include:

- Full name
- Permanent address
- Date of birth and age
- Date of last donation

The donor will be required to provide proof of identification. For the first-time donor, two forms of identification are required.

Age of donor: The minimum acceptable age is 17 or 16 years with parental consent. Donor also needs to provide consent. If the donor is over 75 years of age, he/she may be acceptable at the discretion of the medical director.

Table 1.1 Highlights of Uniform Donor Questionnaire (UDHQ)

Type of Question	Time Frame	Examples (answer to these questions should be "yes" or "no")
General health	Current	Feeling healthy?
		Have a chance to read education materials related to blood donation?
Any recent infection	Current	Taking any antibiotic or being treated for infection?
Recent medication history	Past 2 days	Did you take aspirin or any product containing aspirin?
Long-term medication history	Past 2 days– forever	Taking ceratin medication may disqualify a person from blood donation. Please see Table 1.2
Activity in past weeks	Past 8 weeks	Did you donate blood, platelets or plasma? Did you receive any vaccination or other shots?
		Have contacted anyone who received vaccination for small pox?

Table 1.1 Highlights of Uniform Donor Questionnaire (UDHQ) (*cont.*)

Type of Question	Time Frame	Examples (answer to these questions should be "yes" or "no")
Activity in past weeks	Past 16 weeks	Did you donate a double unit of red cells using an apheresis machine?
Activity in past year	Past 1 year	Did you receive a blood donation?
		Are you an organ transplant recipient?
		Did you receive a bone or skin graft?
		Did you come in direct contact with someone's blood?
		Did you have an accidental needle stick?
		Had sexual contact with a HIV positive or AIDS, patient or a prostitute or any anyone known to abuse illicit drugs?
		For a male did you have sexual contact with a male?
		For a female did have sexual contact with a male who had sexual contact with another male?
		Did you have sexual contact with or live with someone with hepatitis?
		Do you have tattoo, ear or body piercing?
		Did you receive treatment for syphilis or gonorrhea?
		Have you been in juvenile detention, lockup, jail, or prison for more than 72 consecutive hours?
Activity in past years	Past 3 years	Have you ever been outside the United States or Canada?
Specific time frame in past years	1980–1996	Did you spend time that adds up to 5 years or more in United Kingdom?
		Are you in US military or a dependent of a US military personnel?
Long past to present	1980–present	Did you spend time that adds up to 5 years or more in certain European countries or had blood transfusion?
Any time in life	Ever	Are you pregnant or had past pregnancy?
		Have you ever had malaria, Chagas disease or babesiosis?
		Have you ever received a dura mater (or brain covering) graft or xenotransplantation product?
		Did you ever diagnosed with any type of cancer, including leukemia?
		Did you have heart disease, lung disease, or any blood disorder?
		Have any of your relatives had Creutzfeldt–Jakob disease?
		Have you ever tested positive for HIV?
		Have you ever used illicit drug, or work as a prostitute?

AABB website, full questionnaires is available in AABB website.

Table 1.2 Highlights of Abbreviated Donor History Questionnaire (aDHQ) for Returning Blood Donors

Type of Question	Time Frame	Examples (answer to these questions should be "yes" or "no")
General health	Current	Feeling healthy?
		Have a chance to read education materials related to blood donation?
Recent medication history	Past 2 days	Did you take aspirin or any product containing aspirin?
Activity in past weeks	Past 8 weeks	Did you donate blood, platelets or plasma?
		Did you receive any vaccination or other shots?
		Have contacted anyone who received vaccination for small pox?
Activity in past weeks	Past 16 weeks	Did you donate a double unit of red cells using an apheresis machine?
Since last blood donation	Any time after past donation	For a female donor, are you pregnant or had pregnancy since last donation?
		For a female donor, do you have sexual contact with a male who had sexual contact with another male in the past 12 months?
		For any donor, do you live with a person diagnosed with hepatitis or have sex with such person?
		For any donor, did you have any new medical diagnosis, treatment or are you taking any medication that may disqualify you from donation (please see Table 1.3)?
		Have been outside United States or Canada?
		Did you have any accidental needle stick or use needle to inject drugs not prescribed by your doctor?
		Did you get a positive test result for HIV or have sexual contact with a HIV positive person or sexual contact with a prostitute or anyone who use illicit drugs?
		Did you work as a prostitute?
		Did you have sexual contact with or live with someone with hepatitis?
		For a male donor did you have sexual contact with another male?
		Do you have tattoo, ear or body piercing?
		Have you been in juvenile detention, lockup, jail, or prison for more than 72 consecutive hours?
		Have any of your relatives had Creutzfeldt–Jakob disease?

AAABB website at aabb.org.

DONOR EDUCATIONAL MATERIALS AND CONSENT

Donor educational materials provide information about the donation process. It explains the various steps of donor screening including the rationale for the questionnaire. It informs potential donors of behaviors, symptoms, and signs which if present increases possibility of being infected with human immunodeficiency virus (HIV) or having acquired immunodeficiency syndrome (AIDS). In this case, a donor should refrain from blood donation. AABB standards also require that donors should be given educational materials regarding the risks of postdonation iron deficiency. The donor also needs to sign a statement that he or she has read and understood the educational materials. By signing, the donor is certifying that:

- The donor is at least 17 years of age or if younger has parental consent
- Donation is voluntary
- Donor should not donate blood if he/she has AIDS, or tested positive for HIV, or believe that he/she has been exposed to HIV
- The donor blood will be tested for HIV, hepatitis, syphilis, and other infectious agents
- The donor understands that if the blood is positive for any infectious agents, the blood will not be used
- The donor will be informed of any significant positive blood test
- If the donor is deferred then his/her name may be placed on the donor deferral registry that will affect his/her eligibility to donate in the future

BRIEF PHYSICAL EXAMINATION

A potential donor is evaluated on the following parameters:

- *Weight*: Donors should weigh at least 110 lbs. If they weigh less, blood may still be drawn but the amount of blood must be calculated and the anticoagulant in the blood bag must be reduced accordingly. The minimum acceptable weight in this case is 88 lbs. This, however, is rarely done.
- *Temperature*: The body temperature must not exceed 99.5°F or 37.5°C.
- *Pulse*: The pulse rate should be between 50 and 100 per minute with no evidence of irregular rhythm. Lower rates are acceptable from individuals with high exercise tolerance.
- *Blood pressure (BP)*:
  - Systolic BP: between 90 and 180 mmHg
  - Diastolic BP: between 50 and 100 mmHg
- *Hemoglobin level*:
  - *For females*: Hemoglobin levels should be at least 12.5 g/dL (or hematocrit: Hct 38%) for allogeneic whole blood and non-DRBC (double RBC) apheresis donation

- *For males*: Hemoglobin levels should be at least 13 g/dL (or Hct 39%) for allogeneic whole blood and non-DRBC apheresis donation
- *Donating RBC apheresis*: Hemoglobin level should be 13.3 g/dL (Hct 40%)
- Maximum hemoglobin level of 20 g/dL (Hct 60%) if donating whole blood
- Maximum 18 g/dL (Hct 54%) for platelet apheresis, plasmapheresis, and leukapheresis
- Minimum hemoglobin level of 11.0 g/dL for autologous donations
- *Arm inspection*: The phlebotomy site must be clear of any signs of infection or intravenous (IV) drug abuse. If there are signs of IV drug abuse, the donor center should permanently defer the donor

LABORATORY TESTING

Laboratory testing of donor blood is done to ensure that the recipient receives the safest possible blood.

Laboratory testing includes:

- ABO and Rh typing
- Testing for red cell antibodies
- Infectious disease screen for:
  - Human T-lymphotropic virus (HIV-1 and HIV-2)
  - Human T-cell lymphotropic virus (HTLV)-I and HTLV-II (HTLV-I is associated with adult T-cell leukemia-lymphoma and both are associated with HTLV-associated myopathy)
  - Hepatitis B (HBV) and hepatitis C virus (HCV)
  - West Nile virus (WNV)
  - Zika virus
  - *Treponema pallidum* (*T. pallidum*)
  - *Trypanosoma cruzi* (*T. cruzi*)

In addition to these tests, at times some donated units may be tested for *Babesia microti* (*B. microti*) and Cytomegalovirus (CMV) antibodies. All apheresis platelet units are tested for bacterial contamination within 24 h of collection.

METHODOLOGY USED FOR INFECTIOUS DISEASE SCREEN

Nucleic acid testing (NAT) either by polymerase chain reaction (PCR) or transcription-mediated amplification is used to test for HIV-1, HCV, HBV, WNV, and Zika virus. HIV NAT detects HIV-1 RNA which further reduces the window period of infection and positive test result [2]. HCV NAT testing (minipool type) has reduced the window period to about 8–10 days compared to the 70-day window period using HCV enzyme immunoassay (3.0 antibody testing) [3,4]. In the United States, NAT testing for HIV, HBV, and HCV is performed on pools of donor samples ranging from 6 to 16 samples. A triplex assay is used which detects HBV DNA and HIV and

HCV RNA. If there is a positive result, additional testing is done to determine which of the donor-tested positive and for which virus. The time period between infectivity of a blood sample and its detection by minipool testing (MP-NAT) is 9 days for HIV and 8 days for HCV. This results in the risk of HIV or HCV infection from blood transfusion to 1 per 1–2 million.

For WNV, MP-NAT or individual donation NAT (ID-NAT) may be used. It has been demonstrated that WNV RNA positive units may be missed by MP-NAT due to low viral titers. Thus, it is better to use ID-NAT for WNV. ID-NAT is especially pertinent when WNV infection rates rise in certain geographic areas. ID-NAT is also done for Zika virus [5].

IMMUNOASSAYS FOR ANTIBODIES OR ANTIGENS

The immunoassays that are used for such testings are enzyme-linked immunosorbent assay (EIA) or chemiluminescence enzyme immunoassay (CLIA). If a test is negative, then the sample is deemed negative. If a test is positive, then it is labeled as initially reactive. The test is then repeated in duplicate in the same assay system. If one or both of the repeat tests is positive, then the sample is deemed positive. If both the repeat samples are negative, then the sample is deemed as negative. This methodology is used to detect any potential presence of various antibodies including anti-HIV-1 and anti-HIV-2 antibodies, anti-HTLV-I and anti-HTLV-II antibodies, anti-HCV antibody, HBsAg (hepatitis B surface antigen) and anti-HBc antibody, and *T. cruzi* antibody.

If the anti-HIV-1 and anti-HIV-2 antibody tests are positive, then further confirmatory testing are indicated. For HIV-1, this can be done by either Western blot or immunofluorescence assay. For HIV-2, an additional anti-HIV-2 supplemental assay is needed. If the anti-HCV antibody test is positive, then confirmation is achieved by using results of HCV NAT testing or second manufacturer's enzyme or chemiluminescence assay.

If anti-*T. cruzi* antibody test is negative, this needs not to be repeated on subsequent donations because there is very little ongoing transmission of the *T. cruzi* parasite to individuals residing in the United States. If the test is positive, confirmation is done by recombinant antigen-based immunoblot assay. The FDA does not require that if an individual tests positive for anti-HBc for the first time, he/she has to be deferred from donating blood. At least two independent donation tests need to be positive for deferral. However, many donor centers defer individuals if they test positive once.

SEROLOGICAL TESTINGS

Detection of antibodies against *T. pallidum*-specific antigens is used for detection of syphilis. Testing for CMV and *B. microti* is not mandatory. *B. microti* is an obligate intraerythrocytic protozoan that causes babesiosis. Infection in humans is due

Table 1.3 Infectious Disease Testing of Prospective Donors

Infective Agent	Tests
Human immunodeficiency virus (HIV-1 and HIV-2)	All units; anti-HIV-1 and HIV-2 antibody with confirmation if positive; NAT for HIV RNA
HCV	All units; anti-HCV antibody with confirmation if positive; NAT for HCV RNA
HBV	All units; HBsAg and anti-HBc antibody; NAT for HBV DNA
Human T-lymphotropic virus (HTLV I and II)	All units; anti-HTLV antibody
WNV	All units; NAT for WNV RNA
T. pallidum	All units; specific serology for T. pallidum
Zika virus	All units; NAT for viral RNA
T. cruzi	First donation; anti-T. cruzi antibody with confirmation if positive
B. microti	Units in high-risk areas; antibody testing or with PCR
CMV	Selected units for designation of CMV negative units

B. microti, Babesia microti; CMV, *cytomegalovirus;* HBV, *hepatitis B virus;* HCV, *hepatitis C virus;* T. cruzi, Trypanosoma cruzi; T. pallidum, Treponema palladium; WNV, *West Nile virus.*

to infected tick bites. In the United States, the high-risk areas are upper midwest and northeast regions. In these high-risk areas, testing for *B. microti* is indicated. Testing is performed by immunoassays or PCR.

CMV infection can cause severe disease in the immunocompromised patients. CMV safe blood is sought for such patients. Leukodepleted products are considered to be CMV safe. CMV-tested negative is also another source of CMV safe blood. Testing for CMV antibody is done to detect donors who are CMV antibody negative. These individuals have not been exposed to CMV and are CMV-tested negative. A summary of blood bank testing for infectious diseases is available in Table 1.3. The relative risk of acquiring infection from blood products is available from Table 1.4.

Table 1.4 Relative Risk of Acquiring Infection from Whole Blood, Red Cells, Platelets, and Plasma

HIV	1 in 1.5 million
HBV	1 in 1 million
HCV	1 in 1.2 million
HTLV	1 in 2.7 million
Bacterial infection from apheresis platelets	1 in 50,000–80,000

Note: *There is no risk of HTLV infection from plasma.*
HBV, *Hepatitis B virus;* HCV, *hepatitis C virus;* HIV, *human immunodeficiency virus.*

RECIPIENT, DONOR PROTECTION, AND DONOR DEFERRALS

Recipients are protected from infectious agents and noninfectious risks because donors are screened extensively by a combination of educational materials provided to donors, using the UDHQ, miniphysical exam, and laboratory testing.

AVOIDING RISK OF HUMAN IMMUNODEFICIENCY VIRUS INFECTION

Prospective donors are given educational material which explains the risk of HIV transmission and behaviors associated with HIV risk. Donors are instructed not to donate if a risk factor behavior is present. Questions 11–26 and 33–35 in the original UDHQ with the exception of questions 21 and 22 are asked to protect recipients against HIV.

Permanent donor deferral due to HIV risk:

- Previous positive test for HIV
- Ever used nonprescription injection drugs
- Ever having had sex in exchange for money or drugs
  Twelve-month deferral due to HIV risk:
- If the prospective donor had a blood transfusion, from the date of transfusion
- If the prospective donor had a transplant such as organ, tissue, or bone marrow
- If the prospective donor had a graft such as bone or skin
- If the prospective donor had come into contact with someone else's blood
- If the prospective donor had an accidental needle-stick
- If the prospective donor had sexual contact with anyone who has HIV/AIDS or has had a positive test for the HIV/AIDS virus
- If the prospective donor had sexual contact with a prostitute or anyone else who takes money or drugs or other payment for sex
- If the prospective donor had sexual contact with anyone who has ever used needles to take drugs or steroids, or anything not prescribed by their doctor
- If the prospective male donor had sexual contact with another male [6]
- If the prospective female donors had sexual contact with a male who had sexual contact with another male in the past 12 months
- If the prospective donor has a tattoo (need not be deferred if the tattoo was obtained in certified venues)
- If the prospective donor has ear or body piercing (need not be deferred if the piercing was obtained in certified venues)
- If the prospective donor has or been treated for syphilis or gonorrhea
- If the prospective donor had been in juvenile detention, lockup, jail, or prison for more than 72 consecutive hours

If a donor tests positive for HIV, then he/she is permanently deferred.

VIRAL HEPATITIS (HBV AND HCV)

Permanent donor deferral due to hepatitis risk:

- If testing for HBV or HCV yields positive HBsAg, anti-HCV antibody followed by confirmation, positive NAT testing for HBV or HCV, then the donor is permanently deferred. An individual is also permanently deferred, if he/she tests positive for anti-HBc twice.

Twelve-month deferral due to hepatitis risk:

- If the prospective donor has a blood transfusion, from the date of transfusion
- If the prospective donor has a transplant such as organ, tissue, or bone marrow
- If the prospective donor has a graft such as bone or skin
- If the prospective donor has come into contact with someone else's blood
- If the prospective donor has an accidental needle-stick
- If the prospective donor has sexual contact with a person who has hepatitis
- If the prospective donor has lived with a person who has hepatitis
- If the prospective donor has a tattoo (need not be deferred if the tattoo was obtained in certified venues)
- If the prospective donor has ear or body piercing (need not be deferred if the piercing was obtained in certified venues)

WEST NILE VIRUS AND ZIKA VIRUS

Recipients are primarily protected from these agents by means of laboratory testing. Prior to implementation of testing for Zika virus, the FDA recommended a 4-week deferral for individuals who traveled to an area with active Zika transmission, those with symptoms of Zika virus infection and those who had sexual contact with a man who had traveled to or resided in an area with active Zika virus transmission during the prior 3 months. Individuals who test positive for WNV are to be deferred for 120 days.

MALARIA

Transmission of malaria by blood transfusion is rare in the United States [7,8].
    Three-year deferral due to malaria risk:

- If the prospective donor had malaria
- If the prospective donor is an immigrant or resident of malaria endemic countries (defined by residing in that country for more than 5 years)
    One-year deferral due to malaria risk:
- If the prospective donor resides in a country nonendemic for malaria and travels to a malaria-endemic country but did not develop any symptoms related to malaria

CHAGAS DISEASE

- Individuals who respond to having Chagas disease or are tested positive for Chagas disease are permanently deferred.

BABESIOSIS

- Individuals who respond to having Babesiosis are permanently deferred.

CREUTZFELDT–JAKOB DISEASE AND VARIANT CREUTZFELDT–JAKOB DISEASE

Creutzfeldt–Jakob disease (CJD) is a rare, fatal neurodegenerative disease caused by a prion. It has a long asymptomatic latent period. CJD has been transmitted from human to human by transplantation of dura mater and by injection of pituitary-derived human growth hormone. Familial forms of CJD are also known to exist. All of these individuals are deferred permanently from blood donation. Moreover, variant Creutzfeldt–Jakob disease (vCJD) is also a fatal neurodegenerative disorder discovered in the United Kingdom in 1996 [9]. The prion which causes vCJD also causes bovine spongiform encephalopathy (also known as mad cow disease) [10]. Transmission through blood transfusion has been reported on several cases.

Prospective donors are permanently deferred for risk of vCJD, if:

- The prospective donor spent time that adds up to 3 months or more in the United Kingdom from 1980 to 1996
- The prospective donor was a member of the US military, a civilian military employee, or a dependent of a member of the US military, stationed in Europe from 1980 to 1996
- The prospective donor has spent time that adds up to 5 years or more in Europe, from 1980 to present
- The prospective donor has received a blood transfusion in the United Kingdom or France, from 1980 to present
- If the prospective donor has received bovine insulin injection sourced from the United Kingdom or other countries with BSE

BACTERIAL INFECTION

Bacterial infections can cause fatal complications after transfusion. As a result, prospective donors are deferred if:

- They are not feeling healthy and well
- They are taking any antibiotics
- If they are taking any medicine for infections

In addition, the prospective donor's temperature is taken as part of the miniphysical examination.

RECENT VACCINATION

Vaccination with live attenuated viruses has the theoretical risk of transmitting infection. These are the recommendations for live attenuated vaccines:

- Two-week deferral for measles, mumps, oral polio, oral typhoid, and yellow fever
- Four-week deferral for rubella or varicella zoster

NONINFECTIOUS DISEASE RISKS

History of malignancy in donor is important to determine if a donor is eligible or ineligible to donate blood. Following criteria are used:

- If a prospective donor has history of hematologic malignancy, then such donor is permanently deferred from blood donation.
- If the prospective donor has history of solid organ tumor, then such donor should be deferred till the person is considered to be clinically cured (definition varies from 1 to 5 years).
- If a donor has certain malignancies such as basal cell carcinoma of skin or cervical carcinoma in situ, such donor is eligible to donate blood.

MEDICATIONS TAKEN BY DONORS

Potential teratogenic medication taken by a prospective donor is a cause for deferral. Such medications include:

- One month for isotretinoin and finasteride
- Six months for dutasteride
- Three years for acitretin
- Permanent deferral for etretinate

Moreover, if a donor has taken bovine or human growth hormone from the pituitary gland, the donor is permanently deferred from blood donation. In addition, 1-year deferral is also recommended for a donor who has received hepatitis B immune globulin.

If a donor is taking aspirin or anticoagulants, following guidelines are followed:

- If the donor is donating platelets by apheresis, then he/she should not have taken aspirin in the last 48 h or clopidogrel or ticlopidine in the last 2 weeks.
- Prospective donors are temporarily deferred if they are taking warfarin, heparin, or other anticoagulants because use of these drugs causes poor efficacy of the plasma component as well as also increases risk of bleeding for the donor.

Table 1.5 refers to the medications which will result in donor deferral.

DONOR'S PROTECTION

Donors are questioned about their general health as well as specific questions related to lung and heart disease. Another question asks about bleeding disorders, primarily

Table 1.5 List of Medications Which May Lead to Donor Deferral

Medication Use Time Frame	Medication Class/Intended Use	Specific Drugs
Past 2 days	Antiplatelet agent	Piroxicam
	Anticoagulants	Rivaroxaban, dalteparin, enoxaparin, dabigatran, apixaban, edoxaban
Past 7 days	Antiplatelet agent	Prasugrel, ticagrelor
	Anticoagulant	Warfarin, heparin, fondaparinux
Past 14 days	Antiplatelet agent	Clopidogrel, ticlopidine, vorapaxar
Past 1 month	Acne treatment	Isotretinoin
	Used for treating enlarged prostate but also used for preventing scalp hair loss	Finasteride
Past 6 months	Treating prostate problems	Dutasteride
Past 1 year	Hepatitis treatment	Hepatitis B immune globulin
Past 2 years	Treating basal cell skin cancer	Vismodegib
	Treating relapsing multiple sclerosis	Teriflunomide
Past 3 years	Treating psoriasis	Acitretin
Forever	Treating psoriasis	Etretinate
Past 1 year to ever	Miscellaneous	Experimental medication or unlicensed (experimental) vaccine (depending of study, use of certain medications or vaccine may unqualified a donor from blood donation from anytime between past 1 year or ever

Notes: *Insulin from cows (bovine or beef insulin) manufactured in the United Kingdom as well as growth hormone from human pituitary glands are also banned forever drug list for a donor. However, these products are not available in the United States. Donors should not discontinue medications prescribed or recommended by their physician in order to donate blood.*
AAABB website at aabb.org.

to protect the donor from excessive bleeding postdonation. Female donors should not donate till 6 weeks postdelivery.

Presyncopal and syncopal attacks are a common reaction to blood donation. Rates are higher amongst first time and female donors. Donors under the age of 20 also have increased reactions. Consumption of predonation water, eating salty snacks, and distraction techniques with good interaction between donor and phlebotomist may help reduce the rate of such reactions.

DONATION CRITERIA FOR HOMOLOGOUS PLATELET APHERESIS

Following criteria are used for donors in this category of blood donation:

• Donors need to meet usual allogeneic blood donation requirements

- A platelet count is recommended prior to donation. If this is not possible predonation sample needs to be evaluated. Platelet count should be greater than 150,000
- Donor should be off aspirin for 48 h or clopidogrel/ticlopidine for 2 weeks
- Donor is not allowed to donate more than 24 platelet apheresis platelets in a 12-month period
- The interval between two apheresis collections should be at least 2 days
- Maximum of two procedures in a 7-day period

DONATION CRITERIA FOR APHERESIS PLASMA DONATIONS

Following criteria are used for this purpose:

- Donors need to meet usual allogeneic blood donation requirements
- *Occasional plasma donors*: Donor can donate plasma every 4 weeks, to a maximum of 13 donations in a year. The donor is allowed to donate a maximum of 12 L in 12 months (14.4 L, if the donor weighs >175 lbs)
- Frequent plasma donors:
  - Donor has to wait at least 48 h between two procedures
  - No more than two procedures within 7 days
  - RBC volume loss must not exceed 25 mL/week
  - Serum or plasma needs to be tested for total protein and serum protein electrophoresis or quantitative immunoglobulins. Results need to be normal

APPROVED FREQUENCY OF DONATIONS

Following criteria are used to determine frequency of blood donation:

- Donors are eligible to donate whole blood every 8 weeks
- Donor is not allowed to donate more than 24 platelet apheresis platelets in a 12-month period
- Donor can donate plasma every 4 weeks to a maximum of 13 donations in a year
- Donors are allowed to donate DRBC units every 112 days

GRANULOCYTE TRANSFUSION

Granulocytes are collected by apheresis. The donors are premedicated with G-CSF (granulocyte-colony stimulating factor) and steroids. Circulating neutrophil counts in G-CSF and dexamethasone stimulated donors are maximal 12 h after administration. Thus, donors may be given these medications the evening before and collection can be performed the next morning. Some centers choose to administer dexamethasone only where as some centers may avoid both agents. In addition to being qualified as a whole blood donor, donors who intend to donate granulocytes need to meet the following requirements:

- ABO and Rh matched with the recipient
- Donors cannot be pregnant (G-CSF is not approved for pregnancy)

- Donors with hemoglobinopathies
- No history of allergy to steroids or starch (hydroxyethyl starch is added to the donor's blood as it enters the centrifuge to facilitate separation of granulocytes from RBCs)
- Donors should not have medical conditions where steroids are contraindicated (e.g., hypertension, diabetes, GI ulcers, tuberculosis, or fungal infections)

Once the granulocytes are collected, they should ideally be transfused within the next few hours. Granulocytes may be stored at room temperature for 24 h.

DONOR DEFERRAL REGISTRY

When a donor is deferred, his/her name is included in a computerized registry. This registry is used to ensure protection of recipients. Blood centers should check the name of the donor prior to donation. This will also ensure unnecessary phlebotomy of ineligible donors.

KEY POINTS

- A donor can donate: one unit of whole blood, two units of RBCs, platelets apheresis, or plasma apheresis
- There are four major steps of donor selection:

    1. Donor screening by means of a medical history interview
    2. Reading of educational materials
    3. Undergoing a brief physical examination
    4. Laboratory testing

- The minimum acceptable age is 17 or 16 years with parental consent. Donor also needs to provide consent. If the donor is over 75 years of age, he/she may be acceptable at the discretion of the medical director
- Donors should weigh at least 110 lbs
- Donor body temperature must not exceed 99.5°F or 37.5°C
- Donor pulse rate should be between 50 and 100 per minute with no evidence of irregular rhythm. Lower rates are acceptable from individuals with high exercise tolerance
- Donor BP:
    - Systolic BP: between 90 and 180 mmHg
    - Diastolic BP: between 50 and 100 mmHg
- Donor hemoglobin level:
    - *For females*: hemoglobin levels should be at least 12.5 g/dL (or hematocrit: Hct 38%) for allogeneic whole blood and non-DRBC apheresis donation
    - *For males*: hemoglobin levels should be at least 13 g/dL (or Hct 39%) for allogeneic whole blood and non-DRBC apheresis donation
    - *Donating RBC apheresis*: hemoglobin level should be 13.3 g/dL (Hct 40%)
    - Maximum hemoglobin level of 20 g/dL (Hct 60%) if donating whole blood

- Maximum 18 g/dL (Hct 54%) for platelet apheresis, plasmapheresis, and leukapheresis
- Minimum hemoglobin level of 11.0 g/dL for autologous donations
- Donor laboratory testing includes: ABO and Rh typing, testing for red cell antibodies, and infectious disease screen for:
  - Human T-lymphotropic virus (HIV-1 and HIV-2)
  - Human T-cell lymphotropic virus (HTLV)-I and HTLV-II (HTLV-I is associated with adult T-cell leukemia-lymphoma and both are associated with HTLV-associated myopathy)
  - HBV and HCV
  - WNV
  - Zika virus
  - *T. pallidum*
  - *T. cruzi*
- All apheresis platelet units are tested for bacterial contamination within 24 h of collection.
- NAT either by PCR or transcription mediated amplification is used to test for HIV-1, HCV, HBV, WNV, and Zika virus.
- The immunoassays that are used for such testings are EIA or CLIA. This methodology is used for HIV-1 and HIV-2 antibodies, HTLV-I and HTLV-II antibodies, anti-HCV antibody, HBsAg (hepatitis B surface antigen) and anti-HBc antibody, and *T. cruzi* antibody.
- Detection of antibodies against *T. pallidum* specific antigens is used for detection of syphilis.
- Leukodepleted products are considered to be CMV safe. CMV tested negative is also another source of CMV safe blood. Testing for CMV antibody is done to detect donors who are CMV antibody negative.
- Following criteria are used for donors for homologous platelet apheresis:
  - Donors need to meet usual allogeneic blood donation requirements
  - A platelet count is recommended prior to donation. If this is not possible, predonation sample needs to be evaluated. Platelet count should be greater than 150,000
  - Donor should be off aspirin for 48 h or clopidogrel/ticlopidine for 2 weeks
  - Donor is not allowed to donate more than 24 platelet apheresis platelets in a 12-month period
  - The interval between two apheresis collections should be at least 2 days
  - Maximum of two procedures in a 7-day period
- Donation criteria for apheresis plasma donations
  - Donors need to meet usual allogeneic blood donation requirements
  - *Occasional plasma donors*: The donor is allowed to donate a maximum of 12 L in 12 months (14.4 L if the donor weighs >175 lbs)
  - Frequent plasma donors:
    - Donor can donate plasma every 4 weeks, to a maximum of 13 donations in a year

- Donor has to wait at least 48 h between two procedures
- No more than two procedures within 7 days
- RBC volume loss must not exceed 25 mL/week
- Serum or plasma needs to be tested for total protein and serum protein electrophoresis or quantitative immunoglobulins. Results need to be normal
- Following criteria are used to determine frequency of blood donation:
  - Donors are eligible to donate whole blood every 8 weeks
  - Donor is not allowed to donate more than 24 platelet apheresis platelets in a 12-month period
  - Donor can donate plasma every 4 weeks, to a maximum of 13 donations in a year
  - Donors are allowed to donate DRBC units every 112 days

REFERENCES

[1] Zou S, Stramer SL, Dodd RY. Donor testing and risk: current prevalence, incidence, and residual risk of transfusion-transmissible agents in US allogenic donations. Transfus Med Rev 2012;26:119–28.

[2] Glynn SA, Busch MP, Dodd RY, Katz LM, et al. Emerging infectious agents and the nation's blood supply: responding to potential threats in the 21st century. Transfusion 2013;53:438–54.

[3] Busch MP, Kleinman SH, Nemo GJ. Current and emerging infectious risks of blood transfusions. JAMA 2003;289:959–62.

[4] Stramer SL, Glynn SA, Kleinman SH, Strong DM, et al. Detection of HIV-1 and HCV infections among antibody negative donors by nucleic acid-amplification testing. N Engl J Med 2004;351:760–8.

[5] http://www.fda.gov/NewsEvents/Newsroom/PressAnnouncements/ucm518218.htm.

[6] FDA Guidance for Industry: Revised Recommendations for Reducing the Risk of Human Immunodeficiency Virus Transmission by Blood and Blood Products, December 2015. http://www.fda.gov/downloads/BiologicsBloodVaccines/GuidanceComplianceRegulatoryInformation/Guidances/Blood/UCM446580.pdf.

[7] Guerrero IC, Weniger BG, Schultz MG. Transfusion malaria in the United States, 1972–1981. Ann Intern Med 1983;99:221–6.

[8] Nahlen BL, Lobel HO, Cannon SE, Campbell CC. Reassessment of blood donor selection criteria for United States travelers to malarious areas. Transfusion 1991;31:798–804.

[9] Will RG, Ironside JW, Zeidler M, Cousens SN, et al. A new variant of Creutzfeldt–Jakob disease in the UK. Lancet 1996;347:921–5.

[10] Hill AF, Debruslais M, Joiner S, Siddle KC, et al. The same prion strain causes vCJD and BSE. Nature 1997;389:448–50.

Blood components: Processing, characteristics, and modifications

2

INTRODUCTION

This chapter discusses processing, characteristics and modifications of blood products. In general, the processing of blood into components can be performed from either whole blood donations or apheresis methods. However, some of these products are primarily collected from apheresis methods, such as platelets, and some components, specifically granulocytes, are exclusively prepared by apheresis methods. In addition, standards and regulatory measures presented in this chapter are practices performed in the United States to ensure the safety, purity, potency, and quality of blood components. Such safety standards may be different outside the United States.

BLOOD COMPONENT PROCESSING

When whole blood is collected, it must be in a blood container that has been approved by the Food and Drug Administration (FDA) and meeting the requirements of the Code of Federal Regulation Title 21 Part 640. These requirements include a container that is sterile, pyrogen-free, and can be identified by a lot number. Currently, blood containers are made of polyvinylchloride (PVC) as it offers many advantages over glass blood containers that were used more than 50 years ago. Some of the advantages of these containers are the portability and ease of processing as PVC containers are easily malleable and more resilient to damage [1]. Additionally, these materials are permeable allowing for improved gas exchange which is essential for some blood products, such as platelets, and allows for improved storage periods. PVC containers have a certain temperature in which they lose the flexibility and damage resistant properties and become a hard and brittle material, almost like handling a piece of glass. This is known as the glass transition temperature and occurs at temperatures at about $-25°C$ to $-30°C$.

When blood is collected into these containers, current AABB standards allow for the maximum collection of 10.5 mL of blood per kilogram of the donor's weight. The volume of blood collected is dependent on the capacity of bag used, but typically the volume of blood collected is 450 mL ± 10%, equating to 405–495 mL, or 500 mL ± 10%, equating to 450–550 mL of blood collected. Generally speaking, the volume of total blood collected from a whole blood donation can be calculated using the formula for the density of whole blood.

Volume of blood (ml) = weight in grams (g) of unit collected ÷ 1.05 g/ml

If a collection is unable to meet the expected volume, this disrupts the balance between the anticoagulant and whole blood within the bag. Plasma based blood products, such as fresh frozen plasma or platelets, prepared from low volume collections would contain higher amounts of anticoagulation than expected compared to standard collections. Therefore, it is acceptable to collect "low volume units;" however, only red blood cell (RBC) components should be made from these units, and plasma and platelet products should be discarded. A low volume unit is when 300–404 mL of whole blood is collected in a 450-mL bag or 333–449 mL of whole blood is collected in a 500-mL bag.

There are various forms of anticoagulants and preservatives which are used to maintaining stability of blood components and aid in extending the shelf-life of the product (Table 2.1). These products include acid citrate dextrose solution Formula A (ACD-A), acid citrate dextrose solution Formula B (ACD-B), citrate phosphate dextrose solution (CPD), citrate phosphate double dextrose solution (CP2D), and citrate phosphate dextrose adenine solution (CPDA-1). Whole blood may be collected in CPD, CP2D, or CDPA-1, whereas ACD is mainly used as an anticoagulant during apheresis procedures. Both CPD and CP2D allow extension of the shelf-life of RBC components to 21 days while CPDA-1 allows extension of RBC components to 35 days. This is because the addition of adenine to the preservative allows for improved production of adenine triphosphate (ATP) which prolongs RBC survival and thus storage.

Typically, whole blood is collected in either CPD or CP2D. When whole blood is separated into components, the RBC blood component is more viscous with a

Table 2.1 Anticoagulation-Preservative Solutions for Blood Collections

Profile	CPDᵃ	CP2Dᵃ	CPDA-1ᵃ	AS1ᵇ	AS-3ᵇ	AS-5ᵇ	AS-7ᵇ
Shelf-life (days)	21	21	35	42	42	42	42
Contents in mg							
Sodium citrate	1660	1660	1660	0	588	0	0
Citric acid	188	206	188	0	42	0	0
Dextrose	1610	3220	2010	2200	1100	900	1585
Sodium phosphate (monobasic)	140	140	140	0	276	0	0
Adenine	0	0	17.3	27	30	30	27
Mannitol	0	0	0	750	0	525	1000

ᵃBased on 450-mL whole blood collection bag containing 63 mL of anticoagulant/preservative.
ᵇBased on addition of 100 mL of additive solution (AS) to a 450 mL collection bag after soft spin to RBC component. AS-1, AS-5, and AS-7 are added to blood collected in CPD. AS-3 is added to blood collected in CP2D.
Adopted from AABB Technical Manual and Circular of Information.

higher hematocrit due to the loss of anticoagulant/preservatives which are being transferred into plasma-based products. To overcome this problem, additive solution (AS) is routinely added to RBC blood components. AS similar to CPDA contains adenine which improves ATP production allowing for increased RBC viability and a shelf-life of 42 days. Additionally, the addition of AS to RBC units allows for a final hematocrit around 55% allowing for a less viscous and more easily infusible blood product. Currently, there are four FDA-approved ASs: AS-1 (Adsol), AS-3 (Nutricel), AS-5 (Optisol), and AS-7 (SOLX) (Table 2.1). Of the four currently available ASs, only AS-3 lacks mannitol. Mannitol induces diuresis and can affect cerebral fluid dynamic in neonates. However, small volume transfusion (5–15 mL/kg) with RBC units collected in AS is acceptable for neonates including preterm infants.

Once whole blood has been collected, it undergoes centrifugation to allow for separation into components based on their densities. In the United States, one unit of whole blood is initially processed through a soft spin which is a low *g*-force centrifugation that allows separation of the whole blood into RBCs and platelet rich plasma (PRP). After this first centrifugation process, the unit may then be divided into a unit of RBCs and PRP via diversion of the PRP into a satellite bag, typically done using a manual expresser. Noted that the buffy coat or layer of leukocytes between the RBCs and PRP is not as well demarked and residual leukocytes will be in the RBC product as well as the PRP. The PRP in the satellite bag will then undergo a hard spin. This allows formation of the platelet pellet and platelet poor plasma (PPP). It is important to remember that those residual leukocytes in the PRP are now with the platelet pellet. Again, a manual expresser is used to divert the PPP into a satellite bag allowing production of a unit of platelets and a unit of PPP or plasma. The timing for freezing the unit of plasma will determine the designation given to the final product (FFP, plasma frozen within 24 h after phlebotomy, etc.).

Based on the current methods used for separation of blood by centrifugation, it is possible to obtain three major blood components: RBCs, platelets, and plasma. Cryoprecipitated antihemophilic factor (AHF), commonly referred to as "cryoprecipitate or cryo," is a cold insoluble product that precipitates out when FFP is thawed at refrigerator temperatures (1–6°C). The residual plasma is transferred to a satellite bag leaving behind a unit of cryoprecipitate which can be refrozen. The diverted plasma is now relabeled as plasma cryoprecipitate reduced and can be refrozen to be used at a later date.

Labels that are applied to each manufactured blood component are regulated by the FDA and must comply with requirements highlighted in the Guidelines for Uniform Labeling of Blood and Blood Components as well as a FDA-approved bar code system. Most facilities have adopted the International Society of Blood Transfusion 128 bar code symbology as AABB's Standards require facilities to comply with this bar code labeling system for accreditation. Much emphasis is placed on the requirements and standardization of labeling as use of these labeling systems can improve efficiency, traceability, quality, and safety of blood components. The regulation requires that the information on a blood product be presented in a manner that is eye and machine readable and that the label contains at least the following bar code information:

1. Unique facility identifier
2. Lot number relating to the donor
3. Product code
4. ABO group and Rh type of the donor

These requirements apply to all blood and blood components and must be performed by all centers that collect and prepare blood, including hospital transfusion services that issue pooled units and/or issue divided units or aliquots. Additional information may be applied using special labels, such as further product modifications, one example being irradiation [2]. An example of a labeled unit of RBCs is given in Fig. 2.1. Note that the donor identification number contains the facility identification number, collection year period, and a serial number assigned to that specific collection associated with the donor.

BLOOD COMPONENT CHARACTERISTICS

In this section, blood components are discussed.

WHOLE BLOOD

Whole blood can be used for transfusion, although rarely used with the advent of component therapy. Whole blood transfusion is practiced more commonly in the military setting, however there is emerging interest in the civilian trauma setting. Whole blood is typically collected in CPD. Since this product is not divided into components, it has a final hematocrit similar to that of patients, approximately 33%. The shelf-life is 21 days if collected in ACD, CPD, and CP2D and 35 days if collected in CPDA-1. Whole is stored at 1–6 °C. However, for transporting the product, the temperature range is extended to 1–10°C. Characteristics of cellular blood components are summarized in Table 2.2.

RED BLOOD CELLS

As mentioned earlier, whole blood once divided into components leaves a RBC product which is viscous and has a high hematocrit around 65%–85%. The AABB standards require that all RBC products have a final hematocrit less than 80%. ASs are typically added as these preservatives extend the shelf-life of the product to 42 days and lower the hematocrit to approximately 55%. If RBCs were collected by apheresis methods, it is required that at least 95% of units have a hemoglobin greater than 50 g or RBC volume greater than 150 mL. For all RBC units, the amount of hemolysis present at the end of storage needs to be less than 1% with at least 75% of the transfused cells remaining after 24 h proven by in vivo labeling studies.

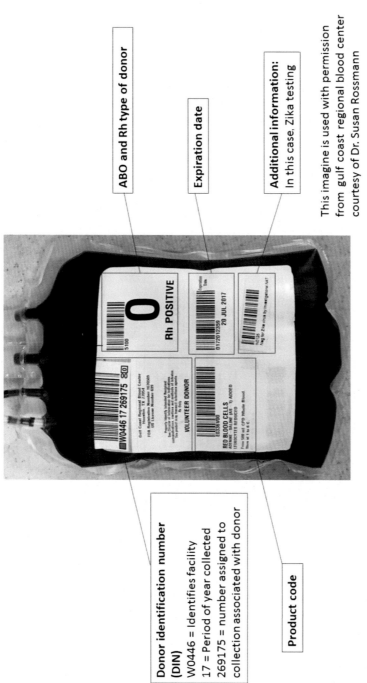

Donor identification number (DIN)

W0446 = Identifies facility

17 = Period of year collected

269175 = number assigned to collection associated with donor

Product code

ABO and Rh type of donor

Expiration date

Additional information: In this case, Zika testing

This imagine is used with permission from gulf coast regional blood center courtesy of Dr. Susan Rossmann

FIGURE 2.1 Labeled RBC unit.

Table 2.2 Cellular Blood Component Characteristics

Product	Expiration	Storage	Transport	Quality Criteria	Comment
Red blood cells					
Whole blood/RBCs	21 days: ACD/CPD/CP2D 35 days: CPDA-1 42 days: AS	1–6°C	1–10°C	End Hct <80% Hemolysis at end of storage should be <1% with retention of 75% of RBC at 24 h	Only RBCs can be made with low volume units. At least 95% of apheresis units have >50 g Hgb or 150 mL RBC volume
Leukoreduced RBCs	21 days: ACD/CPD/CP2D 35 days: CPDA-1 42 days: AS	1–6°C	1–10°C	$<5 \times 10^6$ leukocytes with retention of ≥85% of RBC content in at least 95% of units	At least 95% of apheresis units have Hgb > 42.5 g or 128 mL red cell volume
Irradiated RBCs	Original expiration or 28 days, whichever first	1–6°C	1–10°C	At least 25 Gy to center of bag and 15 Gy to remainder	Max 50 Gy to product
Frozen RBC	10 years	≤−65°C in 40% glycerol	Frozen	Frozen within 6 days of collection, unless rejuvenated, or up to date of expiration if rare unit	Rejuvenated RBCs restores 2,3-DPG and ATP to normal levels
Deglycerolized RBC	24 h if open system 14 days if closed system	1–6°C	1–10°C	Must recover ≥80% RBCs	
Washed RBCs	24 h, open system	1–6°C	1–10°C		
Platelets					
Platelet concentrate	5 days, 4 h if open system, such as in pooling	20–24°C with continued agitation	20–24°C agitation not required	≥5.5×10^{10} platelets with pH ≥6.2 in ≥90% of units	Up to 24 h without agitation for transport

Table 2.2 Cellular Blood Component Characteristics (*cont.*)

Product	Expiration	Storage	Transport	Quality Criteria	Comment
Apheresis platelets	5 days	20–24°C with continued agitation	20–24°C agitation not required	$\geq 3.0 \times 10^{11}$ platelets with pH ≥ 6.2 in $\geq 90\%$ of units	Up to 24 h without agitation for transport
Leukoreduced platelet concentrate	5 days, 4 h if open system, such as in pooling	20–24°C with continued agitation	20–24°C agitation not required	$\geq 95\%$ of units have $< 8.3 \times 10^5$ leukocytes $\geq 75\%$ of units have 5.5×10^{10} platelets and $\geq 90\%$ of units have a pH ≥ 6.2	Up to 24 h without agitation for transport
Leukoreduced apheresis platelets	5 days	20–24°C with continued agitation	20–24°C agitation not required	$\geq 95\%$ of units have $< 5.0 \times 10^6$ leukocytes $\geq 90\%$ of units have $\geq 3.0 \times 10^{11}$ platelets with pH ≥ 6.2	Up to 24 h without agitation for transport
Irradiated platelets	No change in original expiration	20–24°C with continued agitation	20–24°C agitation not required	At least 25 Gy to center of bag and 15 Gy to remained	Max 50 Gy to product Up to 24 h without agitation

Granulocytes

Product	Expiration	Storage	Transport	Quality Criteria	Comment
Apheresis granulocytes	24 h	20–24°C	20–24°C	$\geq 75\%$ of units have 1.0×10^{10} granulocytes	Irradiated Never Leukoreduced

Adopted from AABB Technical Manual and AABB Standards for Blood Banks and Transfusion Services.

These measures are in place as there are physiologic changes that occur within a RBC unit during storage which is termed the "storage lesion." These changes include cell membrane changes such as decreased deformability and increased osmotic fragility, formation of microvesicles, decreased pH, ATP, and 2,3-diphosphoglycerate as well as an increase in lipid peroxidation, potassium, and free hemoglobin.

RBC units are labeled as low volume when containing 300–404 mL of whole blood in a 450-mL bag or 333–449 mL of whole blood in a 500-mL bag. Such blood may only be used for RBC transfusion and cannot be used to make other components.

When RBCs units are released from or received into a transfusion service, a visual inspection of the unit is performed. This is done to identify any abnormalities with the unit such as the presence of hemolysis, clot formation, or color change suggestive of bacterial contamination. If any abnormalities are detected at visual inspection, the RBC unit should not be released from the blood bank [3]. Additionally, RBC units received in a transfusion service have reconfirmation of the ABO type as well as the Rh status if the unit is Rh negative. Weak D testing is not performed.

PLATELETS

Platelets prepared from whole blood are typically suspended in 40–70 mL of plasma. This single component of platelets that is produced from a whole blood donation (whole blood derived platelets) is referred to as a unit of platelets or a platelet concentrate. It is required that at least 90% of units of platelets have at least 5.5×10^{10} platelets and have a pH greater than or equal to 6.2 by the end of its shelf-life. In regards to apheresis platelets, at least 90% units must have at least 3.0×10^{11} and have a pH greater than or equal to 6.2 by the end of the shelf-life. Apheresis platelets are considered a dose of platelets. Typically, six units of platelets, sometimes referred to as a "six pack," are also considered a dose of platelets. This is because when one unit of platelet (5.5×10^{10}) is multiplied by 6, it is equivalent to the required standard for an apheresis platelet (3.0×10^{11}).

Platelets are stored at room temperature between 20°C and 24°C as studies have shown that platelets stored below 20°C have poor in vivo recovery and survival [4]. The shelf-life of platelets is 5 days. The short half-life of platelets is not secondary to functionality as in vitro studies suggest longer shelf-life of 7–10 days. It is the increased risk of bacterial contamination and possibility of a septic transfusion reaction that may occur from platelets stored longer than 5 days at room temperatures that has led to the currently accepted shelf-life of platelets.

During storage platelets undergo continuous agitation. This is because platelets are metabolically active and the continuous agitation allows the platelets to remain suspended in the preservatives as well as gas exchange across the permeable storage bag. While stored, platelets are undergoing glycolysis and oxidative phosphorylation allowing for the production of ATP, thus preserving cellular function. During glycolysis, lactic acid is formed which is buffered by the preservative solution. However, over time, lactic acid can overcome the buffering solution allowing for an acidic pH. This is why, a pH of 6.2 is used as a quality measure during storage as platelets below this pH level have poor in vivo recovery. Additionally, the bag in which platelets are stored allows for gas exchange. This allows entry of oxygen into the bag for oxidative phosphorylation. Moreover, the carbon dioxide produced from oxidative phosphorylation aids in maintaining the bicarbonate buffer, which is ultimately released from the bag. If platelets were to undergo prolong periods without agitation, the metabolic byproducts, mainly lactic acid, would build up and results in damage from the decreased pH [5]. However, studies have shown that platelets show no significant injury if stored up to 24 h without agitation. As a result, platelets are stored with continuous agitation but may go up to 24 h without agitation during transport [6].

GRANULOCYTES

Unlike other blood products, granulocytes are exclusively prepared by apheresis methods. Typically, when granulocytes are collected, they are from donors that have been stimulated with corticosteroids and granulocyte colony stimulating factor as the combination of these medications produces greater mobilization of granulocytes and a better yield from the collection process. By AABB standards 75% of granulocyte collections must have at least 1×10^{10} granulocytes. Additionally, because there is less than ideal separation between the RBC and granulocyte layer, granulocyte collection can often be contaminated with more than 2 mL of RBCs. If red cell contamination is present in more than 2 mL, then the unit should be ABO compatible and Rh matched with the recipient. Better separation of the layer between RBCs and granulocytes can be achieved with the use of hydroxyethyl starch (HES). Once granulocytes have been collected they are stored at 20–24°C for up to 24 h. Additionally, irradiation is done for granulocytes units. One last important point, leukoreduction or use of a microaggregate filter should not be done on this product because such process will removed granulocytes. However, a standard blood filter or macroaggregate filter should be used during transfusion (Table 2.2).

PLASMA

Plasma is the major acellular blood component that contains plasma protein, including all coagulation factors. Major types of plasma products that are available include FFP, plasma frozen within 24 h after phlebotomy (PF24), plasma frozen within 24 h after phlebotomy held at room temperature up to 24 h after phlebotomy (PF24RT24), thawed plasma, plasma cryoprecipitate reduced, and liquid plasma. Coagulation factors present in plasma depend on the method of collection as well time of freezing [7]. Therefore, final label given to the product depends on these factors. It is also important to note that generally for plasma products that are stored frozen, once thawed, these products are good for up to 24 h as the product currently labeled as in refrigerator temperatures (1–6°C). After the 24-h period, these products must be relabeled as thawed plasma (Table 2.3).

Thawed plasma

Thawed plasma is any plasma product (FFP, PF24, PF24RT24) that has been thawed and has been stored longer than 24 h. At that point, the product must be relabeled as thawed plasma. The major difference between thawed plasma and frozen plasma, such as FFP, is that thawed plasma has decreased amounts of cofactors, factor V and factor VIII. However, thawed plasma has acceptable levels of the other factors including ADAMTS 13 (A Disintegrin and metalloproteinase with a thrombospondin type 1 motif, Member 13). Thawed plasma has a shelf-life of 5 days when stored

Table 2.3 Acellular Blood Component Characteristics

Product	Expiration	Storage	Transport	Quality Criteria	Comment
Plasma					
FFP	24 h once thawed and stored at 1–6°C	−18°C for 1 year or −65 °C for 7 years	Frozen or 1–10°C	N/A	Frozen within 8 h, contains approx. 1 IU/mL of clotting factors
PF24	24 h once thawed and stored at 1–6°C	−18°C for 1 year	Frozen or 1–10°C	N/A	Clinically equivalent to FFP
PF24RT24	24 h once thawed and stored at 1–6°C	−18°C for 1 year	Frozen or 1–10°C	N/A	Decreased F8 levels, collected specifically from apheresis methods
Thawed plasma	5 days	1–6°C	1–10°C	N/A	Decreased F5 and F8 levels, expiration date takes into account original day of thawing product
Cryoprecipitate-reduced plasma	24 h once thawed and stored at 1–6°C	−18°C for 1 year	Frozen or 1–10°C	N/A	Indicated for TTP
Cryoprecipitate AHF					
	6 h after thawing 4 h if pooled	−18°C for 1 year	Frozen or 20–24°C	≥80 IU factor VIII and ≥150 mg/dL fibrinogen	Contains also vWF, factor XIII, fibronectin Quality standards applies to all units tested

Adopted from AABB Technical Manual and AABB Standards for Blood Banks and Transfusion Services.

at refrigerator temperatures, 1–6°C, and having this product available is beneficial because it decreases the turnaround time as well as reduce cost and wastage of frozen plasma inventory. Thrombotic thrombocytopenic purpura (TTP) is a good example in which thawed plasma can be used, since the goal of the plasmapheresis in this case is to remove the antibodies to ADAMTS 13 as well as replenish ADAMTS 13. Another important point about thawed plasma, although the shelf-life is 5 days, this includes the original thawing of the product from which it was relabeled from. Example being that if one unit of FFP that is thawed and then after 24 h relabeled as thawed plasma, that thawed plasma product is only good for an additional 4 days. So, the shelf-life of 5 days includes the time in which the product (FFP, PF24, and PF24RT24) was originally thawed.

Fresh frozen plasma
In order for a plasma product to be labeled as FFP, it must be frozen within 8 h of collection. FFP contains all coagulation factors in normal quantities with about 1 international units (IU)/mL of each clotting factor. Note, unlike other blood products, there is no quality measures required to ensure the contents of the product. FFP can be stored frozen for 1 year when frozen at −18°C and 7 years if frozen at −65°C. FFP is typically thawed in a water bath at 30–37°C. Once thawed, the product is considered FFP and stored at refrigerator temperatures 1–6°C for up to 24 h. After this time, the product can be used as thawed plasma which aids in reducing wastage of plasma products.

Plasma frozen within 24 h after phlebotomy (PF24)
PF24 is plasma frozen within 24 of collection, which is logistically more feasible than FFP (which must be frozen within 8 h of collection), particularly for blood donations collected from donor drives. The coagulation factor levels of PF24 have been shown in studies to be within the normal range and show comparable levels to that of FFP. As a result, PF24 is considered clinically equivalent to FFP and can be used interchangeably. PF 24 is stored at −18°C for up to a year and thawed and stored in temperatures similar to FFP.

PF24RT24
PF24RT24 (plasma frozen within 24 h after phlebotomy held at room temperature for up to 24 h) is prepared specifically from apheresis plasma collections unlike FFP and PF24 which may be prepared from either whole blood or apheresis methods. In the case that PF24 is collected by apheresis methods, it is stored in refrigerator temperatures within 8 h of collection and then frozen within 24 h of the collection [8]. Compared to FFP, PF24RT24 has been shown in studies to have up to a 20% reduction in factor VIII.

Cryoprecipitate-reduced plasma
Cryoprecipitate-reduced plasma (CRP) is the plasma product that remains after the manufacture of cryoprecipitate. However, studies have shown that about 200 mg/dL of fibrinogen can be found in CRP. CRP is prepared by a close system allowing for a storage and expiration date similar to FFP and PF24. After the cryoprecipitate has been removed, CRP is refrozen within 24 h and stored at −18°C for up to a year [9]. CRP is rarely used and indicated for transfusion or plasma exchange in patients with TTP, although the evidence is controversial on its effectives.

Liquid plasma

Liquid plasma is derived from the separation of plasma from whole blood at any time during storage and once separated can be stored at 1–6°C for up to 5 days after the whole blood expiration date. This product is primarily used for the manufacturing of other blood products derivatives, such as albumin, intravenous immunoglobulin, and factor concentrates. This process is done for recovered plasma, plasma prepared from donations intended for transfusion, as well as for source plasma, plasma collected by apheresis methods for the purpose of further manufacturing.

Solvent detergent plasma

Solvent detergent plasma is an effective method for preparing pathogen-reduced plasma. Solvent detergent plasma is a product prepared from the pooling of plasma from no more than 2500 donors that is then treated with a solvent (tri-*n*-butyl) and a detergent (Triton X-100) which allows for the inactivation of lipid-enveloped virus. These units are stored at −18°C for up to a year and up to 24 h once thawed. Compared to FFP, it has a 10% reduction in all coagulation factors with the exception of factor VIII which is reduced by 20% and protein S which is decreased by 50%.

CRYOPRECIPITATED ANTIHEMOPHILIC FACTOR

Cryoprecipitated antihemophilic factor or cryoprecipitate is a product that is prepared from FFP. When FFP is allowed to thaw at refrigerator temperatures (1–6°C), a cold insoluble precipitate is formed. The product is then centrifuged and the residual supernatant is transferred to a separate container which becomes cryoprecipitate-reduced plasma. The precipitated product is called cryoprecipitate. Cryoprecipitate must be refrozen within 1 h of removal from the refrigerator when stored at −18°C. This product can be used up to 1 year from the date of the original preparation. Cryoprecipitate contains:

1. Factor VIII
2. Fibrinogen
3. Von Willebrand factor
4. Factor XIII
5. Fibronectin

By quality standards, cryoprecipitate must contain at least 80 IU of factor VIII and 150 mg/dL of fibrinogen in all units tested. However, average fibrinogen content of most preparations is 250 mg/dL [10]. Additionally, units collected from individuals that have blood group A or B have higher levels of factor VIII than those collected from donors with O blood group. This is likely attributed to the relationship between factor VIII and von Willebrand factor (vWF) as it is well known that patients that are type O have lower plasma levels of vWF. Once thawed, cryoprecipitate has a shelf-life of 6 h when stored at 20–24°C. However, this shelf-life is shortened to 4 h, if the sterility of the product is disrupted, such as during pooling, which is common for cryoprecipitate (Table 2.3).

BLOOD COMPONENT MODIFICATION

Various modifications can be done to blood components after processing to include

1. Leukocyte reduction
2. Irradiation
3. Cryopreservation
4. Volume reduction
5. Washing
6. Pooling

In general, modification of blood products will break the sterility/closed system of the product which will reduce the shelf-life to 24 h for products (RBC, FFP) stored at refrigerator temperatures (1–6°C) or 4 h for products (cryo, platelets) stored at room temperature (20–24°C). Additionally, the above mentioned modifications apply only to cellular blood products with the exception of pooling.

LEUKOCYTE REDUCTION

Residual leukocytes can be found in cellular products, such as RBCs and platelets, as demonstrated by the processing of whole blood into components. In the United States, regulator standards require that leukocyte reduced units contain less than 5 million (5×10^6) leukocytes. Units that tend to fail this criteria are usually collected from sickle cell trait donors as the sickle cells cause obstruction of the leukocyte reduction filters and failure of the product to pass through the filter. Those units that are able to be filtered tend to have a higher leukocyte count than what is required by standards. Leukocyte reduction can be done either during prestorage or poststorage. Prestorage is usually accomplished soon after collection during processing within an in-line filter but must be done within 5 days of collection. Poststorage leukocyte reduction is typically performed by the blood bank transfusion service or at the patient bedside. The major advantage of prestorage leukocyte reduction is that it limits the accumulation of cytokines in the product which occurs during storage.

Leukocyte reduction is a valuable blood component modification as there are several adverse events that can happen as a result of residual leukocytes in the product. The major adverse events leukocyte reduction can reduce include:

1. Febrile non-hemolytic transfusion reaction (FNHTR)
2. Human leukocyte antigen (HLA) alloimmunization
3. Cytomegalovirus (CMV) transmission

This is due to reduction of the number leukocytes present, thus decreasing the likelihood of transfused residual leukocytes being stimulated causing cytokine release and a FNHTR. In addition, less exposure to foreign HLA antigens on leukocytes transfused also reduces incidence of HLA alloimmunization and possibly platelet refractoriness. Lastly, CMV remains dormant in leukocytes. As a result, leukocyte reduction also reduces the risk of CMV transmission. In fact, studies have shown that leukoreduced products are equivalent to CMV seronegative product in terms of CMV transmission [11]. It is important to note that FNHTR can occur as a result of passive transfer of accumulated cytokines in the product. This is particularly important in platelet products, as residual leukocytes in platelet products are more efficient in the accumulation of leukocyte-derived cytokines compared to RBC products as they are stored in cold temperatures. As a result, poststorage platelet leukocyte reduction is less effective in deterring FNHTR than poststorage RBC leukocyte reduction. Other potential benefits of leukocyte reduction include reduced mortality in cardiac surgery patient, decreased transmission of other infection agents such as herpes virus and variant Creutzfeldt–Jakob disease, and possibly prevention of transfusion related immunomodulation.

IRRADIATION

Transfusion-associated graft-versus-host disease (TA-GVHD) is a condition that results due to the presence of residual leukocytes found in cellular blood products. In order to prevent TA-GVHD, those recipients at risk must receive irradiate cellular blood products which causes cellular damage of leukocytes through crosslinking of the DNA and prevention of leukocyte replication. At this time, irradiation is the only modification method approved for the prevention of TA-GVHD. Irradiation is not typically done for FFP and cryoprecipitate as they are acellular product due to the nature of their processing as well as the freeze–thaw cycle they undergo without cryopreservation which will result in destruction in any residual cellular component. However, liquid plasma is never frozen. Irradiation may be done with either cesium-137 or cobalt 60; verification of the dose delivered by these methods must be confirmed annually or semiannually, respectively. The appropriate dose of irradiation is 25 Gy (2500 cGy) to the center of the bag and 15 Gy (1500 cGy) to the remainder of the bag. To ensure that the appropriate dose of irradiation has been applied, irradiation sensitive labels are used that change once the appropriate irradiation dose has been achieved.

It is important to note that no more than 50 Gy (5000 cGy) can be given to any portion of the product as this may result in significant damage to RBC products. The process of irradiation results in decreased red cell recovery as well as efflux of potassium from the RBCs. As a result, irradiated blood products have a shelf-life of 28 days or the original expiration date, whichever comes first. Therefore, if a RBC product was appropriately irradiated, but not transfused, that product could be later issued to a patient where irradiation was indicated without additional irradiation as initial irradiation has already caused cellular modification of residual leukocytes and any further additional irradiation may result in further degradation of the product. Additionally, if irradiated RBC products are given to premature infants or if large volumes are given to newborns, then washing or volume reduction may be indicated to remove the excess potassium. The irradiation of platelets does not result in any clinically significant damage to platelet or change in function.

CRYOPRESERVATION

Cryopreservation or freezing is beneficial for preserving rare phenotype units, examples being phenotypes negative for high-frequency antigens or Bombay phenotype. Freezing of RBCs is done with glycerol. Glycerol is taken up into the cells and prevents the osmotic migration of water from the cells as ice crystals are formed. There are both high and low concentrations of glycerol, but typically a high (40%) glycerol concentration is used for freezing. This process must be done within 6 days of collection, unless the RBCs are rejuvenated or in the case of rare units up to the date of expiration without rejuvenation. These units are stored at −65°C for up to 10 years; however, the expiration date can be extended at the discretion of the medical director. Once needed, these units are thawed at 37°C and must undergo a process that effectively removes any residual glycerol. This process of removing glycerol (deglycerolizing) can result in breaking the sterility of the product which will reduce the shelf-life to 24 h. However, there are now closed systems available that allow these products to be stored for up to 14 days. After deglycerolizing RBC, the product must contain greater than 80% of the original RBCs. Cryopreservation of platelets, although possible, is not commonly done. This may due to the fact that in-vivo platelet survival rates are less than 45%. Unlike RBCs, platelets are preserved in dimethyl sulfoxide and stored for up to 2 years.

VOLUME REDUCTION

Volume reduction is a method that allows for the concentration of a cellular blood product (RBCs or platelets) and for the removal of the supernatant. This modification is beneficial in preventing excess volume to those patients who are as risk for volume overload. Usually, excess plasma protein, additives, or potassium are removed and RBCs are hyperconcentrated to a target hematocrit. This process results in breaking the sterility of the product, as a result, RBCs have a shelf-life of 24 h at 1–6°C and platelets have a shelf-life of 4 h at 20–24°C.

WASHING

Washing is done with 1–2 L of saline in an attempt to wash away everything but the cellular blood components. This process results in breaking the sterility of the product, thus reducing shelf-life identical to blood products modified by volume reduction. There are several indications in which washing is indicating

1. severe allergic/anaphylactic transfusion reactions,
2. neonatal alloimmune thrombocytopenia (NAIT) if maternal platelets are used,
3. removal of excess potassium following irradiation,
4. removal of glycerol from frozen RBCs, and
5. patient with T activation.

Washing is beneficial as it removes offending plasma proteins, thus reducing the risk of severe allergic/anaphylactic transfusion reactions. Washing could also be used as an alternative if IgA deficient cellular blood products are not available. Washing can also be used in treating neonates suffering from NAIT. NAIT is due to maternal platelet antibodies, typically human platelet antigen-1A (HPA-1a). Use of maternally washed platelets provides a source of antigen negative platelets that have had removal of the offending antibody to HPA-1a. Washing is also useful in the removal of unwanted substances such as excess potassium accumulated during irradiation and removal of glycerol for frozen RBCs. T-activation is a condition that results from bacteria's removal of *N*-acetylneuraminic acid from the surface of the RBC that results in exposure of a cryptogenic antigen, the T-antigen in a patient. This can lead to binding to a naturally occurring IgM anti-T antibody resulting in hemolysis. By washing, it is possible to remove any residual naturally occurring anti-T antibodies from cellular blood products.

POOLING

Pooling is done for blood products in order to achieve a standard dose. In the case of platelets derived from whole blood donations, 6 units are typically pooled to make a dose of platelets. In the case of cryoprecipitate, typically 10 units of cryoprecipitate are pooled to make one dose. This process can also be done to reconstitute whole blood with the combination of RBC and FFP which is commonly used in the pediatric setting, particularly for exchange transfusion. In these cases, pooling results in breaking of the sterility of product and creating an open system. As a result, these products have a shelf-life of 4 h after pooling as they are stored at 20–24°C.

KEY POINTS

- There are three major blood products that can be made from whole blood donations. These products include RBCs, plasma products (such as FFP), and platelets. Cryoprecipitate is made from the cold insoluble material that is precipitated from the thawing of FFP at 1–6°C. Granulocytes are exclusively produced by apheresis methods.
- RBCs have a shelf-life of 21 days if prepared in ACD, CPD, and CP2D and 35 days if prepared in CPDA-1 and 42 days if AS is added.
- Plasma products are defined by the time in which they are frozen: 8 h for FFP and 24 h for PF24. PF24RT24 is collected exclusively by apheresis methods and may be stored at room temperatures for up to 24 h but should be frozen within 24 h.
- Once a plasma product (FFP, PF24, PF24RT24) is thawed, it is labeled as the intended product which is valid for up to 24 h when refrigerated at 1–6°C. After that time, it must be relabeled as thawed plasma which is good for up to 5 days from the time of thawing the original product.
- Compared to FFP, PF24 is clinically equivalent, but PF24RT24 has decreased factor VIII, and thawed plasma has decreased factor VIII and factor V.

- Cryoprecipitate contains factor VIII, fibrinogen, vWF, factor XIII, and fibronectin.
- Cryoprecipitate has a shelf-life of 6 h, however, it is typically pooled thus breaking the close system of the product which now has a shelf-life of 4 h at room temperatures (20–24°C).
- In general, any product that breaks the sterility/closed system of the product will reduce the shelf-life to 24 h for products stored at refrigerator temperatures (1–6°C) or 4 h if stored at room temperature (20–24°C).
- Leukocyte reduction is beneficial in reducing the incidence of FNHTR, HLA alloimmunization, and CMV transmission.
- Irradiation is the only approved modification to prevent TA-GVHD and must have a dose of 25 Gy (1500 cGy) to the center of the bag and 15 Gy (1500 cGy) to the remainder of the bag to be considered irradiated.
- Irradiated blood products have a shelf-life of 28 days or the original expiration, whichever is first.
- Volume reduction may be indicated for patients at risk for transfusion-associated circulatory overload and for the concentration of RBCs to reach a target hematocrit which is commonly used in pediatric setting.
- Washing is indicated for prevention of severe allergic/anaphylactic transfusion reactions, removal of excess potassium following irradiation, treatment of NAIT if maternal platelets will be transfused, patients with T activation, and deglycerolized RBCs.

REFERENCES

[1] Carmen R. The selection of plastic materials for blood bags. Transfus Med Rev 1993;7(1):1–10.
[2] Food and Drug Administration. Guidelines for uniform labeling of blood and blood components (August 1985). https://www.fda.gov/downloads/BiologicsBloodVaccines/GuidanceComplianceRegulatoryInformation/Guidances/Blood/UCM080974.pdf.
[3] Klein HG, Spahn DR, Carson JL. Red blood cell transfusion in clinical practice. Lancet 2007;370:415–26.
[4] Holme S, Heaton A. In vitro platelet aging at 22°C is reduced compared to in vivo aging at 37°C. Br J Haematol 1995;91:212–8.
[5] Dumont LJ, AuBuchon JP, Gulliksson H, Slichter SJ, et al. In vitro pH effects on in vivo recovery and survival of platelets: an analysis by the BEST Collaborative. Transfusion 2006;46:1300–5.
[6] Gottschall JL, Rzad L, Aster RH. Studies of the minimum temperature at which human platelets can be stored with full maintenance of viability. Transfusion 1986;26:460–2.
[7] Yazer MH, Cortese-Hassett A, Triulzi DJ. Coagulation factor levels in plasma frozen within 24 hours of phlebotomy over 5 days of storage at 1 to 6 degrees C. Transfusion 2008;48(12):2525–30.
[8] Alhumaidan H, Cheves T, Holme S, Sweeney J. Stability of coagulation factors in plasma prepared after a 24 hour room temperature hold. Transfusion 2010;50(9):1934–42.

[9] Smak Gregoor PJH, Harvey MS, Briet E, Brand A. Coagulation parameters of CPD cryo-precipitate-poor plasma after storage at 4 C for 28 days. Transfusion 1993;33:735–8.

[10] Ness PM, Perkins HA. Fibrinogen in cryoprecipitate and its relationship to factor VIII levels. Transfusion 1980;20:93–6.

[11] Bowden RA, Slichter SJ, Sayers M, Weisdorf D, et al. A comparison of filtered leukocyte-reduced and cytomegalovirus (CMV) seronegative blood products for the prevention of transfusion-associated CMV infection after marrow transplant. Blood 1995;86: 3598–602.

Transfusion reactions

INTRODUCTION

Transfusion reactions may be defined as any adverse reaction associated with transfusion of blood products. Blood transfusions are one of the most common procedures for patients in the hospital. Transfusion reactions are the most frequent adverse event associated with administration of blood products. Transfusion reactions may be seen in up to 1% of transfusions. Transfusion reactions can range from mild to life-threatening events. Transfusion reactions can rarely be fatal. The incidence of such fatal reactions varies from 1 in 0.6 million to 2.3 million [1].

Transfusion reactions may be acute or delayed. Acute transfusion reactions are those temporarily associated with the transfusion of a blood product and takes places within 24 h of transfusion. Delayed transfusion reactions take place after 24 h and maybe observed up to 30 days post transfusion.

Transfusion reactions can also be classified based on the etiology. This classification divides transfusion reactions into immunologic and nonimmunologic subtypes. Acute transfusion reactions are listed in Table 3.1 while delayed transfusion reactions are listed in Table 3.2. Transfusion reactions based on etiology are given in Table 3.3.

ACUTE TRANSFUSION REACTIONS

Acute transfusion reactions include acute hemolytic transfusion reaction, allergic and anaphylactic reactions, febrile nonhemolytic transfusion reaction, septic transfusion reactions transfusion associated acute lung injury, and transfusion associated circulatory overload. These acute transfusion reactions are discussed in this section (see Tables 3.4–3.8).

ACUTE HEMOLYTIC TRANSFUSION REACTION

Acute hemolytic transfusion reaction (AHTR) is a potentially fatal transfusion reaction and can be either due to immune or nonimmune mechanisms. Immune mediated acute hemolytic transfusion reactions are typically due to infusion of red blood cells (RBCs) which are hemolyzed by the recipients anti-A, anti-B, or other antibodies. Acute hemolytic transfusion reaction may be precipitated by as little as 5 mL of

Table 3.1 Acute Transfusion Reactions, Approximate Incidence

Type	Incidence (Per 100,000 Units Transfused)
Acute hemolytic	2–8
Anaphylactic	8
Septic	0.03–3.3
TRALI	0.4–1.0
Circulatory overload	10
Febrile nonhemolytic	200 (If leukoreduced); 1000 (if not leukoreduced)
Urticarial	100

Table 3.2 Delayed Transfusion Reactions, Approximate Incidence

Type	Incidence (Per 100,000 Units Transfused)
Delayed hemolytic transfusion reactions	40
Transfusion associated graft versus host disease (TA-GVHD)	Very rare
Posttransfusion purpura (PTP)	1–2

Table 3.3 Transfusion Reaction Based on Etiology

Immunologic	Acute hemolytic
	Delayed hemolytic or delayed serologic
	Febrile nonhemolytic
	Transfusion related acute lung injury
	Allergic (urticarial)
	Anaphylactic
	TA-GVHD
Nonimmunologic, infectious	Septic
	Viral
Nonimmunologic, other	Circulatory overload
	TRALI (lipid activators of neutrophil model)
	Febrile nonhemolytic (due to cytokines)

Table 3.4 Important Transfusion Reactions Associated With Fever

Transfusion Reactions	Comments
Acute hemolytic transfusion reactions	Potentially fatal
TRALI	Potentially fatal
Sepsis	Potentially fatal
FNHTR	Not fatal

Table 3.5 Important Causes of Transfusion Reactions Associated With Shortness of Breath

Transfusion Reactions	Comments
TRALI	Potentially fatal
TACO	Potentially fatal
Allergic/anaphylactic reactions	Potentially fatal
Sepsis	Potentially fatal

Table 3.6 Frequency of Reactions

Common, approximately 1%	
	Urticarial/allergic reactions
	FNHTR
Less common, less than 1%	
	TACO
	TRALI
Rare	
	Anaphylactic reactions
	AHTR
	Sepsis

Table 3.7 Potentially Fatal Transfusion Reactions

Transfusions Reactions	Comments
TRALI	Type of acute transfusion reaction
TACO	Type of acute transfusion reaction
AHTR	Type of acute transfusion reaction
Sepsis	Type of acute transfusion reaction
Anaphylaxis	Type of acute transfusion reaction
TAGVHD	Type of delayed transfusion reaction

Table 3.8 TRALI Versus TACO

Feature	TACO	TRALI
Temperature	Unchanged	Fever
BP	Hypertension	Hypotension
Neck veins	Engorged	Unchanged
Dyspnea	Present	Present
Chest X-ray	Diffuse bilateral infiltrate	Diffuse bilateral infiltrate
Pulm arterial occlusion pressure	Greater than 18 mmHg	Less than 18 mmHg
BNP	Significantly raised	Not raised
Ejection fraction	Reduced	Normal
Respond to diuretics	Positive	Not so

incompatible RBCs. Transfusion with significant amount of incompatible plasma is another example of such mistransfusion. Rarely, acute immune mediated hemolytic transfusion reaction may be due to transfusion of incompatible plasma, usually from an apheresis platelet transfusion. Nonimmune mediated acute hemolytic reactions may be due to coadministration of RBCs with an incompatible crystalloid solution (e.g., 5% dextrose), incorrectly stored blood or use of malfunctioning or nonvalidated administration system [2,3].

Features of acute hemolytic transfusion reaction include:

- Fever or chills (most common symptom) and maybe at times the only feature
- Flank pain (due to renal capsular distension)
- Hypotension
- Dyspnea
- Hemoglobinuria, hemoglobinemia
- Disseminated intravascular coagulation (DIC)
- Acute renal failure (ARF)
- Shock.

Sources of mistransfusion leading to acute hemolytic transfusion reaction include:

- Wrong blood in tube: improper identification of patient during initial blood sampling
- Testing wrong sample
- Switching compatibility labels
- Errors in typing
- Grabbing wrong unit from shelf
- Misidentification of the patient or blood product at the time of transfusion (most common type of error).

Approximately 70% of such errors are preanalytical errors which occur outside the blood bank.

There are certain forms of hemolytic anemia where hemolysis is exacerbated with transfusion. This increase in hemolysis may mimic acute hemolytic transfusion reaction. These include:

- Warm autoimmune hemolytic anemia (WAHA)
- Cold hemagglutinin disease (CHAD)
- Paroxysmal nocturnal hemoglobinuria (PNH)
- Drug induced hemolysis.

Pathophysiology and management of acute hemolytic transfusion reaction

When the antibody present in a recipient binds with red cells of a donor, complement activation takes place. As a result of complement activation red cells are lysed releasing hemoglobin into the circulation. Mast cell mediators are also released causing vasodilatation and hypotension. The clotting pathway is activated causing disseminated intravascular coagulation (DIC), which results in generation and deposition of fibrin, leading to microvascular thrombi in various organs and contributing to multiple organ dysfunction. Intravascular clotting, hypotension and precipitation of hemoglobin in the renal tubules also results in acute renal failure.

Laboratory findings in acute hemolytic transfusion reaction include:

- Decrease in hemoglobin and or hematocrit
- Hemoglobinemia
- Hemoglobinuria
- Increased lactate dehydrogenase (LDH)
- Increased indirect bilirubin
- Decreased haptoglobin
- Evidence of DIC: prolonged PT, PTT, decreased fibrinogen, increased D-Dimer
- Schistocytes or spherocytes on peripheral smear

Management of acute hemolytic transfusion reaction includes:

- Supportive therapy using IV fluids, pressors
- Maintaining adequate renal perfusion (diuretics to ensure urine output > 1mL/kg/h; dialysis if necessary)
- Management of bleeding from DIC using appropriate transfusion of blood components

ALLERGIC AND ANAPHYLACTIC REACTIONS

Allergic transfusion reactions typically occur during the transfusion or within 4 h of transfusion. Allergic reactions are due to antigens in the donor plasma. Allergic reactions are most frequently associated with platelet transfusions [4]. IgE antibodies present in the recipient bind to antigens from the donor, the reverse is also possible. Symptoms are due to release of histamine from mast cells or basophils. Allergic reactions are common and can occur in 1–3 percent of cases. Most reactions are mild with rash, itching and urticaria. In contrast, anaphylactic reactions can be life-threatening and patients may develop bronchospasm, respiratory distress and hypotension. Anaphylactic reactions can be seen in IgA deficient recipients who have anti-IgA antibodies in their plasma, and in individuals with anhaptoglobinemia who develop anti haptoglobin antibodies. The recipient's anti-IgA antibodies react with IgA antibodies in the donor's blood product. Severe laryngeal edema or bronchospasm has been described in patients undergoing plasma exchanges due to hypersensitivity reactions to ethylene oxide or other agents used to sterilize the apheresis kits [5].

Individuals with allergic reactions can be treated with antihistamines. If the symptoms abate the transfusion may be resumed with the same product. Anaphylactic reactions require intramuscular epinephrine. In addition, antihistamine, IV steroids, and bronchodilators may be used.

Patients with history of mild allergic reactions are not required to be premedicated. However, if history of allergic reaction is more than mild, premedication with antihistamines may be used. Washing the unit prior to transfusion is another way to reduce allergic reactions in individuals with history of moderate to severe allergic reactions. Individuals who are IgA deficient should ideally receive blood from another IgA deficient donor.

FEBRILE NONHEMOLYTIC TRANSFUSION REACTION

Febrile nonhemolytic transfusion reaction (FNHTR) (1%–3% per unit transfused) is most common among all acute transfusion reactions [6]. This type of transfusion reaction may occur from platelets and RBCs but most likely with platelets derived from whole blood. FNHTRs are due to cytokines or recipient antibodies interacting with donor antigens. Implicated cytokines are interleukins (IL-1, IL-6, and IL-8) and tumor necrosis factor-alpha (TNF-alpha). Antibodies against class I HLA antigens or less commonly antibodies to platelets and granulocytes may also be responsible for FNHTR. FNTHRs associated with platelet transfusions have been associated with release of CD154 from platelets which induces generation of cytokines.

In FNHTR the body temperature is greater than 38°C (100.4°F) with a rise in body temperature by 1°C (1.8°F) or greater from the normal body temperature. Chills and rigor may accompany the fever. These findings are seen typically within 4 h of transfusion. Transfusion should be discontinued and a transfusion reaction workup should be initiated. Fever is an important sign for hemolytic transfusion reactions, bacterial contamination and transfusion associated acute lung injury. The unit should be cultured if the rise in temperature is 2°C (3.6°F) or higher. The patient's blood may also be sent for cultures. Antipyretics should reduce the fever. Prestorage leukoreduction significantly reduces febrile nonhemolytic transfusion reactions. Premedication with antipyretics should be discouraged. It does not seem to reduce rate of this type of reaction. In addition, it may mask fever which is a sign for other important transfusion reactions such as hemolytic transfusion reactions. Please note FNHTR is a diagnosis of exclusion. Similar reaction characterized by chills and/or rigor can occur without rise in body temperature with platelet transfusions.

SEPTIC TRANSFUSION REACTIONS

Septic transfusion reactions usually occur within 4 h of transfusion. The most common cause is bacterial contamination of platelet products. The incidence is approximately 1 in 3000–5000 units of platelets transfused. This is primarily because platelets are stored at room temperature. The source of bacteria for platelets is usually from the skin flora at the time of donation. The implicated organisms are typically Gram positive; most commonly *Staphylococcus aureus* followed by *Staphylococcus epidermidis*.

RBC units may also be a potential source of bacterial sepsis. This is commonly from donors who have asymptomatic bacteremia. The implicated organisms are typically Gram negative (*Yersinia enterocolitica* > *Pseudomonas* > *Serratia*).

Gram positive organisms are more likely to cause reactions, but Gram negative organisms are more likely to cause fatality due to release of endotoxins. The usual clinical features are fever, chills, rigor and hypotension. Both the unit and patient should be cultured. Transmission of *Babesia microti* is an important cause of fatalities associated with septic transfusion reactions.

TRANSFUSION-ASSOCIATED ACUTE LUNG INJURY

Transfusion-associated acute lung injury (TRALI) is a life-threatening form of non-cardiogenic pulmonary edema. It is defined as new lung injury during or within 6 h of transfusion. Patients develop hypoxemia and chest X-ray reveals bilateral lung infiltrates. There is no evidence of circulatory overload. Thus, BNP levels should not be raised. The estimated risk of TRALI is 0.4 per 100,000 units of plasma, 1 per 100,000 per apheresis platelets and 0.5 per 100,000 units of RBCs. TRALI is the leading cause of transfusion related mortality in the United States. Due to the risk mitigation strategies plasma components and apheresis platelets do not confer the highest risk for TRALI. Rather, at present it is the transfusion of red cells which has the highest risk of TRALI.

Anti-HLA or antihuman neutrophil antigen (anti-HNA) antibodies are sufficient to cause TRALI (immune mechanism). Most cases, however, are thought to occur through a two-event model (nonimmune mechanism). The first event is a clinical situation which leads to activation of the pulmonary endothelium. This results in sequestration and priming of neutrophils in the lungs. With subsequent transfusion the neutrophils are activated which causes pulmonary endothelial damage with resultant pulmonary edema. The transfused product may be the source of antibodies or pro-inflammatory mediators.

Certain risk factors have been identified for development of TRALI:

- Liver transplantation surgery
- Chronic alcohol abuse
- Shock
- Smokers
- High IL-8 levels
- Positive fluid balance
- Female sex of the donor
- Increased parity of the donor.

Management of TRALI includes:

- Respiratory support (supplemental oxygen, or ventilation if needed)
- Diuretics are not indicated.

Approximately 80% of patients improve within the next 2–3 days. TRALI is now the leading cause of transfusion-related mortality. One established way to reduce the risk of TRALI is to avoid females as a source of donor plasma or apheresis platelets and/or screening them for HLA/HNA antibodies. In case of TRALI, the recipient should not receive any blood products from the same donor in the future.

TRANSFUSION-ASSOCIATED CIRCULATORY OVERLOAD

Transfusion-associated circulatory overload (TACO) is a form of pulmonary edema due to transfusion of an excessive quantity of blood products or an excessive rate of transfusion. It takes place during transfusion or within the next 4–6 h. Features of volume overload are present. Patient develops features of respiratory distress due to pulmonary edema, which are hypoxemia and tachypnea with bilateral basilar crepitations. Chest X-ray also reveals features of pulmonary edema. The brain natriuretic peptide level is often elevated in TACO.

DELAYED TRANSFUSION REACTIONS

Transfusion reactions can also be delayed. Such transfusion reactions are discussed in this section.

DELAYED HEMOLYTIC OR DELAYED SEROLOGIC TRANSFUSION REACTIONS

The incidence of delayed hemolytic transfusion reaction is approximately 1 in 2500 transfusions. This number is significantly increased in patients with sickle cell disease to approximately 11% [7]. Delayed transfusion or delayed serologic reactions occur in recipients with antibodies against red cell antigens which have been acquired through previous transfusion or pregnancy. Delayed serologic transfusion reaction is one in which the patient lacks the clinical features but exhibits the laboratory features. These antibodies with time decrease to undetectable levels. It has been shown that up to 25% of red cell alloantibodies become undetectable about 10 months after the initial development of the antibodies. These individuals are at risk for delayed hemolytic or serologic transfusion reactions. Antibodies to antigens belonging to the Rh, Kell, Duffy, Kidd, MNS, and Diego blood group systems are most often implicated [8]. Antibodies to Kidd blood group system are most often implicated, followed by Duffy and Kell. When a recipient has an underlying undetectable antibody and is transfused with red cells with the same antigen–antibody titers may rise within 24 h to 28 days. The hemoglobin may fall or fail to be elevated, post transfusion. Moreover, indirect bilirubin level in serum is also increased. The direct Coombs test becomes positive. Elution studies when performed should be able to demonstrate the antibody. The clinical features that may be seen include:

- Fever with or without chills
- Jaundice
- Dark urine
- Abdominal or back pain
- Dyspnea
- Hypertension

Examination of the peripheral smear will demonstrate presence of spherocytes. Most patients do not require treatment. Additional transfusions may be required to achieve desired hemoglobin.

Patients with sickle cell disease or thalassemia should receive blood matched at a minimum for Rh (D, C, c, E, and e) and K antigens. The more antigen matched the units are the less likelihood of delayed hemolytic or serologic transfusion reactions.

TRANSFUSION-ASSOCIATED GRAFT VERSUS HOST DISEASE

Transfusion-associated graft versus host disease (TAGVHD) is a very rare adverse reaction but has a very high mortality rate. It is due to transfusion of cellular components which contain viable lymphocytes that recognize the host as foreign. Transfusion with whole blood, packed red cells, platelets, and granulocytes have all been implicated in TAGVHD. Following patients are at risk for TAGVHD:

- Recipients of hematopoietic stem cell transplant
- Congenital immunodeficiency affecting T cells
- Hodgkin lymphoma
- Neonatal exchange transfusions
- Patients on high dose chemotherapy or radiotherapy
- Patients taking purine analog drugs, alemtuzumab or antithymocyte globulin
- Fetuses requiring intrauterine transfusions
- Immunocompetent recipients, receiving blood from first degree relatives.

  Clinical features of TAGVHD include:

- Rash
- Abdominal pain, nausea, vomiting, and diarrhea
- Fever
- Bone marrow failure.

Initial features are observed within 5–10 days. Bone marrow failure usually develops within three weeks. Leukoreduction is not considered to prevent TAGVHD. Cellular components need to be irradiated prior to transfusion, in those patients deemed at risk.

POSTTRANSFUSION PURPURA

Posttransfusion purpura (PTP) is seen as significant thrombocytopenia which develops 1–2 weeks post transfusion with either platelets or RBCs transfusion. Significant thrombocytopenia may result in bleeding and intracranial bleeding may lead to death. PTP usually occurs in HPA-1a (Human platelet antigen) negative individuals who have been previously alloimmunized by pregnancy. Antibodies destroy the HPA-1a antigen positive donor platelets as well as the HPA-1a negative patient platelets. The mechanism by which the patient's own platelets are destroyed is unknown. Diagnosis is confirmed by detecting platelet specific antibodies. Thrombocytopenia may persist for about a month. Management is supportive. Platelet transfusion may not help. Even transfusion with antigen negative donors during the acute phase has not been shown to be effective. Treatment with intravenous immunoglobulins and steroids may be considered. Therapeutic plasma exchange is another consideration. For future transfusions washed red cells or platelets form HPA compatible donors are required.

OTHER TRANSFUSION REACTIONS

Miscellaneous transfusion reactions are discussed in this section.

HYPERHEMOLYTIC TRANSFUSION REACTIONS

These are rare but life-threatening transfusion reactions that typically occur in recipients with hemoglobinopathies (e.g., sickle cell disease). Hyperhemolytic reactions may be acute or delayed. The acute form occurs with 7 days of RBC transfusion. The delayed form takes place after 7 days. Features of hyperhemolysis syndrome include:

- Fall in hemoglobin level
- Rise in indirect bilirubin
- Rise in LDH levels
- Fall in absolute reticulocyte count

Further, transfusion may worsen hemolysis. However, severe anemia may warrant transfusion. In severe cases, intravenous immunoglobulins, steroids may be of use. Rituximab and plasma exchange may also be helpful [9].

TRANSFUSION-ASSOCIATED NECROTIZING ENTEROCOLITIS

This type of transfusion reaction is seen in preterm and very low birth weight infants. The pathogenesis is unknown.

HYPOTENSIVE TRANSFUSION REACTIONS

Acute hypotensive transfusion reactions are uncommon but in this type of transfusion reaction there is an abrupt drop of systolic or diastolic blood pressure by 30 mmHg or more within 15 min of starting the transfusion. The hypotensive episode resolves within a few minutes of stopping the transfusion. In addition to hypotension, respiratory, gastrointestinal, or mild allergic symptoms may also be seen. The mechanism is thought to be due to activation of the intrinsic contact activation pathway with generation of bradykinin [10]. Bradykinins cause drop in blood pressure which results in tachycardia and induce contraction of smooth muscles of the gut. This type of reaction is more often seen in individuals taking angiotensin-converting enzyme inhibitors (ACE inhibitors).

MASSIVE TRANSFUSION-ASSOCIATED REACTIONS

With massive transfusions the following are established side effects:

- Hypocalcemia (due to large volume of citrate)
- Hyperkalemia (leakage of potassium from red cells in stored blood)
- Hypothermia

BLOOD BANK WORKUP FOR TRANSFUSION REACTIONS

When a transfusion reaction workup is initiated by the clinical team, the unit is sent to the blood bank. Pertinent history regarding the nature of the reaction is also provided to the blood bank. At the blood bank, the following steps are taken:

- Clerical check of the component, its label, paperwork, and initial patient sample used for typing and cross matching
- Repeat ABO testing on the posttransfusion patient sample
- A visual check of pre and posttransfusion sample for evidence of hemolysis
- A direct Coombs test on the posttransfusion sample.

KEY POINTS

- Acute hemolytic transfusion reaction may be precipitated by as little as 5 mL of incompatible RBCs.
- DIC and ARF are important manifestations of acute hemolytic transfusion reactions.
- Allergic transfusion reactions typically occur during the transfusion or within 4 h of transfusion. Allergic reactions are due to antigens in the donor plasma. Allergic reactions are most frequently associated with platelet transfusions.
- Anaphylactic reactions can be seen in IgA deficient recipients who have anti-IgA antibodies in their plasma, and in individuals with anhaptoglobinemia who develop anti-haptoglobin antibodies.
- Febrile nonhemolytic transfusion reactions (1%–3% per unit transfused) is most common among all acute transfusion reactions. This type of transfusion reaction may occur from platelets and RBCs but most likely with platelets derived from whole blood. Febrile non-hemolytic transfusion reactions are due to cytokines or recipient antibodies interacting with donor antigens.
- In FNHTR, the body temperature is greater than 38°C (100.4°F) with a rise in body temperature by 1°C (1.8°F) or greater from the normal body temperature. Chills and rigor may accompany the fever. These findings are seen typically within 4 h of transfusion.
- Fever is an important sign for hemolytic transfusion reactions, bacterial contamination and transfusion associated acute lung injury.

- Septic transfusion reactions usually occur within 4 h of transfusion. The most common cause is bacterial contamination of platelet products. The implicated organisms are typically Gram positive; most commonly *Staphylococcus aureus* followed by *Staphylococcus epidermidis*.
- RBC units may also be a potential source of bacterial sepsis. This is commonly from donors who have asymptomatic bacteremia. The implicated organisms are typically Gram negative (*Yersinia enterocolitica* > *Pseudomonas* > *Serratia*).
- TRALI is now the leading cause of transfusion-related mortality.
- TRALI is a life-threatening form of noncardiogenic pulmonary edema. It is defined as new lung injury during or within 6 h of transfusion.
- Anti-HLA or antihuman neutrophil antigen (anti-HNA) antibodies are sufficient to cause TRALI (immune mechanism). Most cases however, are thought to occur through a two-event model (nonimmune mechanism).
- TACO is a form of pulmonary edema due to transfusion of an excessive quantity of blood products or an excessive rate of transfusion. It takes place during transfusion or within the next 4–6 h. The brain natriuretic peptide level is often elevated in TACO.
- Delayed transfusion or delayed serologic reactions occur in recipients with antibodies against red cell antigens which have been acquired through previous transfusion or pregnancy. Delayed serologic transfusion reaction is one in which the patient lacks the clinical features but exhibits the laboratory features Antibodies to antigens belonging to the Rh, Kell, Duffy, Kidd, MNS, and Diego blood group systems are most often implicated.
- TAGVHD is a very rare adverse reaction but has a very high mortality rate. It is due to transfusion of cellular components which contain viable lymphocytes that recognize the host as foreign. Transfusion with whole blood, packed red cells, platelets and granulocytes have all been implicated in TAGVHD.
- Posttransfusion purpura (PTP) is seen as significant thrombocytopenia which develops 1–2 weeks posttransfusion with either platelets or RBCs transfusion.
- PTP usually occurs in HPA-1a (human platelet antigen) negative individuals who have been previously alloimmunized by pregnancy. Antibodies destroy the HPA-1a antigen positive donor platelets as well as the HPA1a negative patient platelets.
- Hyperhemolytic transfusion reactions are rare but life-threatening transfusion reactions that typically occur in recipients with hemoglobinopathies (e.g., sickle cell disease). Hyperhemolytic reactions may be acute or delayed. The acute form occurs with 7 days of RBC transfusion. The delayed form takes place after 7 days.
- Acute hypotensive transfusion reactions are uncommon but in this type of transfusion reaction there is an abrupt drop of systolic or diastolic blood pressure by 30 mmHg or more within 15 min of starting the transfusion. The hypotensive episode resolves within a few minutes of stopping the transfusion.
- Bradykinins cause drop in blood pressure which results in tachycardia and induce contraction of smooth muscles of the gut. This type of reaction is more

often seen in individuals taking angiotensin-converting enzyme inhibitors (ACE inhibitors).
- With massive transfusions the following are established side effects:
- Hypocalcemia (due to large volume of citrate)
- Hyperkalemia (leakage of potassium from red cells in stored blood)
- Hypothermia

REFERENCES

[1] Vamvakas EC, Blajchman MA. Transfusion-related mortality: the ongoing risks of allogeneic blood transfusion and available strategies for their prevention. Blood 2009;113:3406.
[2] Strautz RL, Nelson JM, Meyer EA, Shulman IA, et al. Compatibility of ADSOL stored red cells with intravenous solutions. Am J Emerg Med 1989;7:162–4.
[3] McCullough J, Polesky HF, Nelson C, Hoof T. Iatrogenic hemolysis: a complication of blood warmed by a microwave device. Anesth Analg 1972;51:102–6.
[4] Harvey AR, Basavaraju SV, Chung KW, Kuehnert MJ. Transfusion related adverse reactions reported to the National Healthcare Safety Network Hemovigilance Module, United States, 2010–2012. Transfusion 2014; published online Nov 5, DOI:10.1111/trf.12918
[5] Leitman SF, Boltansky H, Alter HJ, Frederick CP, et al. Allergic reactions in healthy platelet-pheresis donors caused by sensitization to ethylene oxide gas. N Engl J Med 1986;315:1192.
[6] Sanders RP, Maddirala SD, Geiger TL, Pounds S, et al. Premedication with acetaminophen or diphenhydramine for transfusion with leucoreduced blood products in children. Br J Haematol 2005;130:781–7.
[7] Talano JA, Hillery CA, Gottschall JL, Baylerian DM, et al. Delayed hemolytic transfusion reaction/hyperhemolysis syndrome in children with sickle cell disease. Pediatrics 2003;111:e661–5.
[8] Ness PM, Shirey RS, Thoman SK, Busk SA, et al. The differentiation of delayed serologic and delayed hemolytic transfusion reactions: incidence, long term serologic findings, and clinical significance. Transfusion 1990;30:688–93.
[9] Uhlmann EJ, Shenoy S, Goodnough LT. Successful treatment of recurrent hyperhemolysis syndrome with immunosuppression and plasma to red blood cell exchange transfusion. Transfusion 2014;54:384–8.
[10] Moreau ME, Thibault L, Desormeaux A, Chagnon M, et al. Generation of kinins during preparation and storage of whole blood derived platelet concentrates. Transfusion 2007;47:410–20.

Blood bank testing

INTRODUCTION

This chapter addresses steps that are taken by the blood bank to ensure the delivery of blood products to patients in a safe and efficient manner. The discussion includes common aspects to consider that may cause delay in receiving blood products to patients as well as highlight the important pitfall that may result in the transfusion of incompatible units. This chapter also discusses the major serology tests performed in the blood bank as well as additional testing that may be performed in the blood bank for the identification of antibodies.

PRETRANSFUSION TESTING

Pretransfusion testing includes the process of obtaining blood for patients requiring transfusion as well as the serology testing performed in the blood bank. The steps in place are performed to ensure the safest and most efficient practice is followed in transfusing blood products. The required processes for pretransfusion testing are provided by American Association for Blood Banks (AABB) in the *Standards for Blood Banks and Transfusion Services* [1]. These required steps include:

1. Proper patient identification and appropriate collection of blood sample
2. Blood bank evaluation/requirement of specimen and retention of sample
3. Serology testing to include ABO and D type
4. Serology testing for identifying unexpected, clinically significant non-ABO antibodies (screen)
5. Selection of appropriate blood component(s) for the patient
6. Crossmatching of the selected component(s) with the intended patient
7. Labeling and issuing of blood component(s) from the blood bank

REQUEST FOR BLOOD PRODUCTS

When a transfusion is needed for a patient, a request for blood products must be generated by a physician or other qualified health care professional. This order is typically generated electronically in most institutions and must contain sufficient information to appropriately identify the patient requiring transfusion. This is accomplished by using two independent patient identifiers, typically the patient's name and unique medical record number (MRN). Other important information for a blood product request include the product needed (such as RBCs, platelets or plasma), number of units required, the urgency of the need (e.g., routine, STAT, surgery), and any modifications (such as leukocyte reduction, irradiation, sickle cell negative).

PROPER PATIENT IDENTIFICATION AND APPROPRIATE COLLECTION OF BLOOD SAMPLE

Once a request for blood products has been generated, a patient sample should be collected for testing if such specimen is not already available. Before collection of the sample, the patient should be identified with at least two unique identifiers and the collected specimen must be properly labeled at the patient's bedside by the health care professional collecting the specimen. The required information on specimen label includes:

1. Patient identification with two unique identifiers (typically name and MRN)
2. Date of collection
3. Phlebotomist collecting the sample

However, the date of collection and person collecting the sample do not necessarily have to be labeled on the specimen as long as there is a process in place to identify this information for any given blood bank sample.

The patient identification step is critical in ensuring safe transfusion practices as inaccurate identification of a patient may lead to a mistransfusion which may even be fatal. A mistransfusion refers typically to the transfusion of ABO-incompatible RBC units. This can happen as a pre-analytical error when WBIT (wrong blood in tube) takes place. The definition of WBIT includes [2]:

1. Blood is taken from a wrong patient but is subsequently labeled with information of the intended patient requiring transfusion.
2. Blood is collected from the intended patient for transfusion but labeled with another patient's information.

For example, two trauma patients have arrived at the same time to the hospital with patient X having blood type A but patient Y having blood type O. If blood specimens are drawn at the same time from both patients but the labels are reversed (WBIT), then the blood bank workup would incorrectly show patient X with blood type O and patient Y with blood type A. If there is a historical record of the patient's blood type, this error will be recognized. However, if there is no historical type available and if type specific blood is then issued for these individuals, patient Y has the risk of having a transfusion reaction because the patient will receive type A blood which is ABO-incompatible because the patient is type O. It is estimated that the risk of a WBIT is 1 in 2000 specimens [3] and one method to overcome this potential error is to have a blood bank policy that requires a second sample to be collected in addition to the original specimen if there is no historical blood type.

BLOOD BANK EVALUATION/REQUIREMENT OF SPECIMEN AND RETENTION OF SPECIMENS

Once the sample is received in blood bank, the blood bank staff must compare the identifying information on the requisition with the information on the sample label and ensure that all the information is complete, accurate, and legible. If there is any question, then ideally a new sample should be collected. Similar to the strict criteria for the appropriate collection of a pretransfusion sample, the blood bank process is stringent because of the risk of potential errors and a mistransfusion associated with poorly labeled specimens. Lumadue et al. demonstrated that specimens that failed to meet criteria for specimen acceptance were 40 times more likely to have a blood group discrepancy [4]. After the blood bank staff has ensured that the specimen is properly labeled, the blood bank personnel should also review any previous blood bank records for that patient. This will include a review of the followings:

1. Previous ABO and Rh type (can easily identify WBIT at this point if historical record and current sample type do not match)
2. Any difficulties in typing the patient
3. Any history of clinically significant antibodies
4. Any previous transfusion reactions
5. Any special transfusion requirements (Hgb S negative, irradiated, etc.)

Serology testing, discussed in the next section, will be performed on the current blood bank sample and those results will be compared with the previous records. If there are any discrepancies noted, the blood bank staff should undertake further investigation and any necessary action before issuing units from the blood bank. The type of collection tube that is acceptable for the blood bank testing is either a pink top EDTA tube for plasma or a red top tube for serum.

Patient samples that are processed through the blood bank must be retained for at least 7 days and are stored at refrigerator temperatures. This is done to allow for later testing should the patient develop an adverse reaction to transfusion. Additionally, a segment of any RBC containing unit that leaves the blood bank will be retained and stored in the blood bank should later testing be required.

Typically most transfusion services require a new pretransfusion sample every 3 days. This is required when the patient has been recently transfused or pregnant in the last 3 months, or the history is unknown. For the 3-day rule, the day of collection is day zero with most transfusion services having the pretransfusion sample expiring on day 3 at midnight. For example, if a specimen was collected on Monday at 2:30 pm, then the specimen would no longer be valid for crossmatching on Friday at 12:00 am but would have been valid for Thursday at 11:59 pm.

It is also important to understand that crossmatching in regards to the 3-day rule is done for blood products that contain more than 2 mL of RBCs (RBC units, whole blood, and typically granulocytes). This is because of the potential for the patient to form antibodies (alloantibodies) to foreign RBC antigens most commonly through transfusion or fetal–maternal hemorrhage during pregnancy. Thus a sample is required every 3 days in the blood bank for determination of emerging alloantibodies and crossmatch compatibility with RBC containing blood products. In contrast, transfusion of non RBC containing blood products (plasma, platelets, cryoprecipitate) do not require a new pretransfusion sample every 3 days as long as the patient has an ABO/Rh type (type) on the current admission as the transfusion of these blood product are ensuring only ABO compatibility.

SEROLOGY TESTING

The primary tests performed in the blood bank are the type and screen. The blood type includes determination of the patient's ABO antigens and ABO antibodies (isohemagglutinins) and as well Rh status, specifically the D antigen status. The screen is a panel that contains the major clinically significant RBC antigens and is used to look for the presence or absence of alloantibodies and if positive further testing is done to determine the specificity of the alloantibody or alloantibodies.

The ABO Type includes evaluation of ABO antigens (A, B, AB, O) as well as determination of the ABO isohemagglutinins (anti-A, anti-B). Major concept in blood banking is that an individual has the potential to form antibodies to antigens they lack. In the case of isohemagglutinins, these antibodies are naturally occurring and do not require previous exposure of RBC antigens. This is due to the fact that these antibodies are typically produced from environmental exposure with microorganisms that are antigenically similar to the ABO antigens. Therefore, individuals of all blood types will express isohemagglutinins to the antigens they lack (Table 4.1).

These isohemagglutinins are primarily IgM and capable of inducing intravascular hemolysis via activation of classic complement pathway. As a result the ABO antigens and antibodies are considered the most clinically significant blood group. As a result, safety measures such as pretransfusion testing are performed primarily to avoid a mistransfusion involving ABO-incompatible blood to patients. In fact, proficiency testing monitoring as defined by the Joint Commission and the Clinical Laboratory Improvement Amendments (CLIA) require 100% proficiency testing for ABO group and D typing, whereas all other laboratory proficiency testing require at least a score of 80%.

ABO typing includes determining the forward and reverse reaction for a patient's sample:

- In the forward typing the patient's RBCs are tested to determine which ABO antigens are present
- In the reverse typing the patient's plasma/serum is tested to determine which ABO isohemagglutinins are present

Table 4.1 ABO Antigens and Corresponding Antibodies

Blood Type	A	B	AB	O
Isohemagglutinins	Anti-B	Anti-A	None	Anti-A, Anti-B, and Anti-A,B

Table 4.2 Forward Typing Results

	Patient RBCs Mixed with	
Blood Type	Anti-A	Anti-B
O (no A or B antigens)	Negative	Negative
A (A antigens)	Positive	Negative
B (B antigens)	Negative	Positive
AB (A and B antigens)	Positive	Positive

In forward typing, RBCs from a patient are mixed with known antisera (anti-A is stored in blue color coded bottle while anti-B is kept in yellow color coded bottle). Agglutination (positive reaction) after mixing with specific antisera indicates the presence of such antigen on patient's RBC while no agglutination (negative reaction) indicates absence of that antigen (Table 4.2).

In reverse testing, patient's serum is mixed with known antigen expression (A1 and B cells) reagents. For example, an individual with A blood type should lack B antigen and serum collected from this person should agglutinate with B cells due to the presence of naturally occurring anti B antibody in individual's serum. However, serum of such person should show negative reaction when mixed with A1 cells because this antigen is present on their RBCs. Similarly, a person with B blood type (lacking A antigen) will not show agglutination when mixed with B cells but will show positive reaction when mixed with A cells. Individuals with O blood type lack both A and B antigens. Therefore, serum collected from an individual with O blood type should show positive response with both A1 and B cells because these individuals have three natural occurring antibodies in their serum; anti-A, anti-B and anti-A,B. However, people with AB blood type will not react with either A1 or B cells because they have both A and B antigen on their red cells (Table 4.3).

The results of the forward and reverse typing should match for a specific ABO blood type. If not, further bank blood investigation should be done to determine the cause of the discrepancy. If blood products are needed before the discrepancy can be determined, universal components should be issued, type O RBCs and type AB plasma.

Table 4.3 Reverse Typing Results

	Patient Serum Mixed with	
Blood Type	A1 Cells	B Cells
O (anti-A, anti-B present)	Positive	Positive
A (anti-B present)	Negative	Positive
B (anti-A present)	Positive	Negative
AB (no isohemagglutinins present)	Negative	Negative

Common Causes for ABO discrepancies:

1. Increased reactivity on forward typing
   a. Acquired antigens (Acquired B phenotype or B(A) phenomenon)
   b. ABO mismatched from stem cell transplant
   c. Nonspecific agglutination
   d. Out of group transfusion with mixed field
2. Increased reactivity on reverse typing
   a. Cold reacting auto or alloantibodies
   b. Passive transfuse of antibodies such as from out of group plasma transfusion or infusion of IVIG
   c. A2 subgroup patient with A1 antibodies
   d. Increased serum protein
3. Decreased reactivity on forward typing
   a. Decreased antigen expression due to hematologic malignancy
   b. Massive transfusion (e.g., A red cells in AB patient)
   c. Weak ABO subgroup
   d. Newborns with weak ABO expression
4. Decreased reactivity on reverse typing
   a. Immunosuppressed, transplant patients
   b. Elderly and newborns
   c. Hypogammaglobulinemia

D typing is performed to determine the presence or absence of the D antigen on the patient's RBCs. Therefore, testing for the D antigen is similar to forward typing in which the patient's RBCs are mixed with anti-D antisera and a positive or negative reaction indicates either Rh+ or Rh−, respectively. The D antigen is determined in the type because after the ABO antigens, the D antigen is the most clinically significant and immunogenic antigen with approximately 20% of D negative individuals forming an anti-D when exposed to D positive RBCs—RhD alloimmunization [5]. Unlike ABO isohemagglutinins that are naturally occurring, anti-D antibodies require previous exposure typically either through previous transfusion or pregnancy. If there are difficulties in interpreting the D antigen status, for example, a female of child bearing age, the patient should be given Rh negative RBC containing blood products. Additional testing to determine if an individual is weak D should be performed on newborns and donor units. Weak D testing on donor units is performed by the donor center. Transfusion services are required to confirm the Rh negative status of received RBC units; however, testing for weak D is not required by the transfusion services.

Another exception to the normal practice of ABO/Rh typing for neonates includes only performing forward and D typing. The rationale is that isohemagglutinins begin to develop around 4–6 months of age. Therefore, only one determination of the ABO and Rh type is required during the hospital admission or until the neonate has reached the age of 4 months. Reverse typing is only performed if a nongroup-O neonate will receive RBCs other than type O that is not compatible with the mother's ABO type.

Table 4.4 Selection of RBC Containing Blood Products

Patient's Type	Appropriate Blood Product Type[a]
O	O
A	A, O
B	B, O
AB	A, B, AB, O

[a] Whole blood should be type specific.

SELECTION OF APPROPRIATE BLOOD COMPONENT FOR PATIENTS

In regard to ABO compatibility for selection of blood components, RBC containing products (RBCs, whole blood, granulocytes) selected for transfusion should be compatible with the patient's plasma/isohemagglutinins. However, whole blood should be type specific as whole blood contains a significant amount of plasma (Table 4.4). Transfusion of plasma based products (plasma, cryoprecipitate, platelets) should have compatibility of the product's isohemagglutinins with the patient's RBCs (Table 4.5). In regards to platelet transfusion, in addition to ensuring compatibly of the plasma it is also important to remember that platelets contain ABO antigens and that transfusion of ABO identical platelet products will provide the highest correct count increment (CCI) in contrast to ABO incompatible platelet transfusions. For example, it is not uncommon to transfuse type A platelet into a type O patient. In this case the plasma portion of the transfusion is compatible. However, there may be a less than expected increase in the platelet count given that the transfused platelets are incompatible with the patient's isohemagglutinins.

With regards to D compatibility, only RBC blood products (RBCs, whole blood and granulocytes) and to some extent platelets, as platelet products may contain residual RBCs, should be matched for the D antigen to prevent RhD alloimmunization. However, in certain situations it may not be feasible to provide Rh matched products due to inventory issues or in the case of a patient requiring excessive Rh negative units thus causing depletion of the blood bank inventory of the hospital. In these situations it is important to discuss with the clinical team and weigh the risk/benefit of transfusion of Rh positive units. In cases where a Rh negative female receives Rh positive units, RhIg should be given to prevent RhD alloimmunization. It is also important to note that when a Rh negative patient receives a large volume of

Table 4.5 Selection of Plasma Containing Blood Products

Patient's Type	Appropriate Blood Product Type
O	O, A, B, AB
A	A, AB
B	B, AB
AB	AB

Rh positive RBC units, administration of RhIg in this setting should be reconsidered as the dose required to prevent Rh alloimmunization may induce significant hemolysis in the patient. An alternative more acceptable option in this setting is to perform a RBC exchange to prevent RhD alloimmunization. Use of RBC exchange in the prevention of RhD alloimmunization after exposure to Rh+ RBCs is a category III indication according to the American Society for Apheresis (ASFA) guidelines [6].

In addition to ABO and Rh compatibility, selection of appropriate units includes providing RBCs lacking antigens to alloantibodies the patient is known to have a history of or for alloantibodies identified on the current workup. In cases of known or newly identified alloantibodies, a crossmatch, specifically an antihuman globulin (AHG) crossmatch is performed.

CROSSMATCHING OF THE SELECTED COMPONENT WITH THE INTENDED PATIENT

The purpose of the crossmatch is to demonstrate compatibility of the selected RBC blood product with the patient's plasma (major crossmatch) or compatibility of the selected plasma blood product with the patient's RBCs (minor crossmatch) which serves as an additional safety layer to ensure that the appropriate blood products have been selected. If no alloantibodies are identified on the current workup and the patient does not have a previous history of alloantibodies, then ensuring ABO compatibility is the only requirement at the time of crossmatching. ABO compatibility can be demonstrated by either an immediate spin crossmatch or an electronic crossmatch.

Immediate spin crossmatch is a serology test that includes physically testing the patient's sample with the selected blood product and demonstrating no agglutination indicates appropriate units are selected. Electronic crossmatching requires the use of a validated computer system that uses logic to detect ABO incompatibility and alert the users of any discrepancies. The advantages of electronic crossmatch are the decreased sample requirement, improved utilization of the blood bank inventory, as well as the decreased work load on staff and improved turnaround time for issuing blood products from the blood bank.

If the patient has a history of alloantibodies and/or the current workup shows alloantibodies, then an AHG crossmatch must be performed. However, AHG crossmatching is only required for the transfusion of RBC containing blood products that have been selected based on the lack of antigens to the patient's known or newly identified alloantibodies. For this crossmatch, the patient's plasma and selected donor's RBCs are incubated with the addition of an AHG reagent. This type of crossmatching ensures both ABO compatibility as well as compatibility for alloantibodies. If the blood product is compatible, then no agglutination should be observed.

LABELING AND ISSUING OF BLOOD COMPONENTS

At the time of issuing a tag or label containing the patient's two independent identifiers, donation identification number, and interpretation of compatibility results must be attached to the blood container. Additionally, a final check will be performed by the blood bank staff before issuing blood products. This includes verifying:

1. The patient's two independent identifiers, ABO and Rh type
2. Donation identification number, donor unit ABO and Rh type
3. Interpretation of crossmatch results
4. Any special transfusion requirements
5. Expiration date and time of product
6. Date and time of issuing

ANTIBODY IDENTIFICATION

The antibody screen is used to determine the presence of unexpected non-ABO antibodies, alloantibodies. Alloantibodies are broadly divided into clinically significant alloantibodies as well as nonclinically significant alloantibodies. Clinically significant alloantibodies are typically IgG antibodies containing two identical light chains and two identical heavy chains (approximate molecular weight 150 kDa) forming a Y shaped structure. Formation of these antibodies requires previous exposure and these antibodies react at 37°C. In contrast, nonclinically significant alloantibodies are typically IgM antibodies that are similar in structure to IgG antibodies. However they are arranged as a pentamer linked together by a j chain. Nonclinically significant alloantibodies are naturally occurring and react at colder temperatures.

Clinically significant alloantibodies are formed in response to previous sensitization to foreign RBC antigens either through pregnancy or transfusion. It is estimated that approximately 1.2%–35% of the population are sensitized with the development of alloantibodies, alloimmunization. The variability in the degree of alloimmunization is attributed to the age (elderly and pediatric patients are less likely than adults), patient population (patients on immunosuppressive therapy are less likely to develop alloantibodies than sickle cell patients) as well as the number of transfusions. Clinically significant alloantibodies can result in decreased RBC survival with either intravascular or more commonly extravascular hemolysis. This can results in hemolysis in adults as well as hemolytic disease of the fetus/newborn (HDFN).

The Food and Drug Administration (FDA) requires the antibody screen at a minimum to evaluate alloantibodies to the following common RBC antigens:

Rh antigens (D, C, E, c, and e)
Kell antigens (K and k)
Duffy antigens (Fy^a and Fy^b)
Kidd antigens (Jk^a and Jk^b)
Lewis antigens (Le^a and Le^b)
P antigen (P1)
MNS antigens (M, N, S, and s)

The antibody screen uses group O reagent RBCs so that there is no interference from ABO antibodies. Typically two to three reagent cells are used in a screen and includes evaluation of at least the FDA required antigens which are indicated

Table 4.6 Antibody Screen

	Rh					Kell		Duffy		Kidd		Lewis		P1	MNS			
Cell	D	C	E	c	e	K	k	Fy (a)	Fy (b)	Jk (a)	Jk (b)	Le (a)	Le (b)	P1	M	N	S	s
1	+	+	0	0	+	+	+	+	0	0	+	+	0	0	+	+	0	+
2	+	0	+	+	0	0	+	0	+	+	0	0	+	+	0	+	+	0

as being positive or negative on a given reagent RBC by either Plus (+) or Minus (−) sign respectively (Table 4.6). If the antibody screen is found to be positive then an antibody identification panel (an extended panel contains typically 11 different reagent RBCs) is used to determine the specificity of the alloantibody or alloantibodies causing a positive screen. If the screen is negative, then the blood bank can proceed with selection of only ABO compatible units as long as the patient has no previous history of alloantibodies. If an alloantibody is identified and/or the patient has a previous history of alloantibodies, then the blood bank must select RBC units negative for the antigen(s) the patient has formed alloantibodies against.

INDIRECT ANTIGLOBULIN TEST

The principle test used for the antibody screen and antibody identification panel is an indirect antiglobulin test (IAT) (Fig. 4.1). There are a variety of methods that can be used to identify alloantibodies by this testing principle such as gel, solid phase, and tube method. The IAT result is considered positive if agglutination is observed. For this test, the patient's plasma/serum is mixed with reagent RBCs (screen cells). If the patient has alloantibodies present in their serum/plasma, then these antibodies will bind to those reagent RBCs that contain the corresponding antigen—sensitization. However, as mentioned previously, the majority of clinically significant antibodies, with exception being ABO antibodies, are of the IgG class. Therefore, sensitization of these antibodies to the reagent RBCs may not allow for physical agglutination. This is why an AHG reagent is added; to allow for bridging of the IgG coated RBCs and gross visualization of agglutination. Note that for tube method and solid phase method, a wash step is required after incubation of the patient's plasma with the reagent RBCs to remove any unbound antibodies before addition of the AHG reagent.

METHODS FOR ANTIBODY IDENTIFICATION

Tube method includes three phases of testing: Immediate spin (IS), Incubation (enhancement) phase at 37°C, and an AHG (Coombs phase). The IS phase includes mixture of two drops of the patient's plasma/serum with one drop of 2%–5% reagent cells and then centrifugation is performed for approximately 15–30 s. This phase is ideal for identifying IgM antibodies as their pentamer structure will allow for

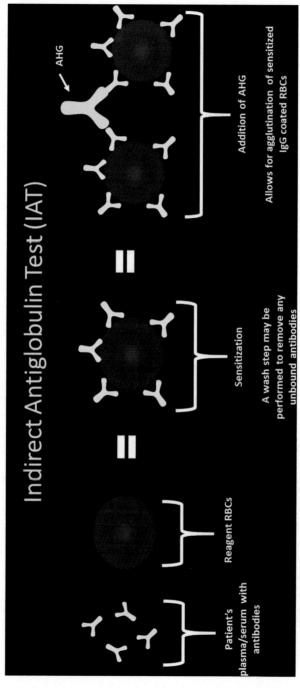

FIGURE 4.1 Principles of indirect antiglobulin test.

visualization of agglutination at this phase. As a result ABO antibodies can be easily detected at this phase and in fact this is an option that can be performed for cross-match ABO compatibility testing. However, other than ABO antibodies, this phase of testing will primarily identify more cold reacting nonclinically significant antibodies. The incubation phase at 37°C allows for sensitization of IgG clinically significant antibodies to their target RBC antigen. However, due to the size of IgG antibodies as well as the zeta potential, the repulsive force created by the negative charge on the RBC surface, agglutination is typically not seen at this phase. Enhancement media can be added during the incubation phase to improve sensitization. Low ionic strength saline (LISS) is one common enhancement media that lowers the zeta potential and increases sensitization. Polyethylene glycol (PEG) is another enhancement media that functions by removing excess water allowing for an increase in the concentration of antibodies and increased sensitization. 22% bovine albumin is another enhancement media that works similar to LISS by lowering the zeta potential. LISS and PEG reduce the incubation phase to about 15 min while 22% bovine albumin reduces the incubation phase to approximately 30 min. It is also important to note that these enhancement media have different effects of warm and cold autoantibodies. LISS and 22% bovine albumin enhance the reactivity of cold autoantibodies while PEG enhances the reactivity of warm autoantibodies. In tube testing after an incubation phase has been performed, the sample is then washed to remove any unbound antibodies before the addition of the AHG (Coombs) reagent. As illustrated previously, the addition of the AHG reagent will allow for agglutination of IgG coated RBCs. Therefore, a positive reaction seen at the AHG phase represents RBCs coated by IgG antibodies that are likely to be clinically significant. The degree of agglutination is rated from 1+ to 4+ with a 1+ reaction being fine granular material seen in the tube and a 4+ reaction being a large single agglutinin identified. When performing tube testing, any reagent RBC line that demonstrates a negative reaction after the addition of AHG must be verified with check cells. Check cells are pre-sensitized cells that are then incubated with the AHG reagent. This step ensures that the AHG reagent is working properly and that the negative reaction seen on testing is valid. If the check cells do not work properly, then the antibody testing is considered invalid and test must be repeated.

Gel method utilizes cards that contain a gel matrix impregnated with AHG. In general, 0.8% of reagent cells plus patient's serum/plasma is then added to a microtubule in the gel card. The sample is then incubated from 15 to 60 min at 37°C depending on what enhancement media if any is used. After incubation, the gel card is then centrifuged and interpreted for the degree of agglutination. In this case, the gel matrix acts as a physical barrier preventing the passage of large agglutinins. In addition, immunogenically, the AHG reagent in the gel matrix will bind to any IgG sensitized RBCs with the degree of agglutination being interpreted as 1+ to 4+. Therefore, a 4+ reaction would cause agglutination toward the top portion of the microtubule while a negative reaction would cause a pellet at the bottom of the microtubule.

The solid phase method utilizes a microwell plate that has been coated with RBC antigens. The patient's sample is then added to the microwells and incubated at 37°C.

A washed step is then performed to remove any unbound antibodies. Indicator RBCs bound to anti-IgG are added to each well and then centrifuged. A positive reaction is determined by the degree in which a diffuse pattern is seen on the bottom of the microwell plate. Therefore, a 4+ reaction would be indicated by a diffuse red pattern seen throughout the entire bottom of the microwell while a negative reaction would be indicated by the presence of a central pellet.

RULES TO ANTIBODY IDENTIFICATION

The first step to antibody identification is the "rule out" step. During this step it is possible to rule out antigens for any RBC reagent cell that did not show reactivity on the panel. This is performed since the alloantibodies that the patient has does not react with the antigens found on a specific reagent RBC. Therefore, one can "rule out" these antigens. Once this rule out process is performed then the next step is to "rule in." During this step it is possible to identify which remaining antigens that were not ruled out fit the pattern of the reactivity seen in the panel.

Once an antibody is identified that fits the pattern seen on the panel then the "rule of 3" should be applied. This is performed to statistically prove that the specificity of the antibody identified on the panel is correct. It is important to note that the alloantibodies identified on the panel will be retained permanently in blood bank records and honored when providing safe compatible blood products. Therefore one should be confident in the alloantibodies identified as this will alter all future transfuse requirements. Using the "rule of 3" one should demonstrate that panels cells negative for the antigen of the suspected alloantibody show no agglutination (negative) while panels cells positive for the antigen of the suspected alloantibody show agglutination (positive). Therefore, demonstrating 3 positive and 3 negative for a suspected alloantibody will meet statistically significance confirming the identity of the antibody identified. While the 3 in and 3 out rule is common, there are variations on the number of positive and negative that can be used to achieve statistical significance, for example, 5 positive and 2 negative. An additional step to verify the alloantibody identified is to phenotype the patient's RBC for the antigen they have developed an alloantibody against. A patient will only form an alloantibody if he/she does not possess the antigen. In this case, the patient should be negative for the antigen confirming their capability of forming alloantibodies to that antigen. Note, phenotyping can only be performed if the patient has not been recently transfused in the last 3 months.

Dosage

Dosage is evaluated when doing the rule out step. Dosage effect refers to the intensity of reactivity seen on the panel for certain RBCs antigens, most notably for Rh, Kidd, Duffy, and MNS blood groups. Alloantibodies against these blood group antigens will react more strongly with those reagents RBCs that are homozygous for a given antigen, while those that are heterozygous will shows a decreased reaction or possibly no reaction at all. Therefore, as a rule, one should not rule out on heterozygous cells that express dosage effect. Note the example of an alloantibody to Jk (a) antigen

Table 4.7 Alloantibody to Jk (a) Antigen Showing Dosage

RBC Phenotype	AHG	Check Cells
Jk (a+b−)	3+	
Jk (a+b+)	1+	
Jk (a−b+)	0	✓
Jk (a+b+)	0	✓
Jk (a+b−)	3+	

(Table 4.7). In this example we see that reagent cells that are homozygous to the Jk (a) antigen express a stronger reaction, while those that are heterozygous express a weak or no reaction at all. Therefore, the rule out step should be performed using homozygous cells but not using the heterozygous cells.

Neutralization

Neutralization can be done to aid in the confirmation of an antibody or can be used to neutralize a certain antibody to allow for investigation of other potential antibodies. Neutralization is performed with specific substances that will inhibit the reactivity of a specific antibody thus neutralizing that antibody's reactivity on the panel. Some common neutralization substances include:

- Saliva from an individual that has the secretor gene to neutralize ABO antibodies
- Saliva from an individual that has the secretor and Lewis gene to neutralize Lewis antibodies
- Hydatid cyst fluid or pigeon egg whites to neutralize P1 antibodies
- Human urine to neutralize Sd (a) antibodies
- Human serum to neutralize Chido/Rodgers antibodies

Enzyme Treatment

Various proteolytic enzymes, most commonly ficin, papain, and bromelin, can be used to enhance or destroy the activity of an antibody's reaction. This technique is useful to enhance the detection/confirmation of an antibody or in its removal to allow for detection of other antibodies. Enzyme treatment enhances the reactivity of alloantibodies to the Rh and Kidd antigens as well as antibodies to carbohydrate antigens (ABO, H, I, P, and Lewis). Enzyme treatment most commonly destroys reactivity to the MNS and Duffy antigens. The Kell blood group antigens are the most notable blood group that shows no effect to enzyme treatment, neither enhanced nor destroyed. A sulfhydryl reagent, such as Dithiothreitol (DTT) or 2-mercaptoethanol (2-ME), must be used to remove the antibody reactivity seen against the Kell antigens.

Lectins

Lectins are seed extracts that show reactivity toward specific RBC antigens. These substances are useful in determining if an individual lacks a common antigen. For example, the lectin *Dolichos biflorus* shows reactivity toward the A1 antigen and *Ulex europaeus* shows reactivity toward the H antigen. In addition, for the identification of universally expressed cryptogenic antigens, (such as the T antigen that may become expressed during infections) the lectin *Arachis hypogea* may be used.

AUTOCONTROL AND DIRECT ANTIGLOBULIN TEST

The autocontrol (AC) is performed when the antibody screen is found to be positive and run in parallel with the extended antibody panel under the same testing method. However, the autocontrol is performed using the patient's RBCs (instead of the reagent RBCs) plus the patient's plasma/serum. A positive AC could represent foreign RBCs coated by alloantibodies, autoantibodies, or antibodies to the enhancement media. Therefore, a thorough investigation of the patient's previous transfusion history should be reviewed. In addition, determination of the urgency for transfusion should be evaluated as the presence of autoantibodies will mask the identification of underlying alloantibodies requiring additional testing to determine the presence of underlying antibodies or additional time providing fully phenotype matched units. While this clinical investigation is conducted a direct antiglobulin test (DAT) should be performed.

The DAT is blood bank test that will determine if there is an immunologic cause for the reactivity seen in the AC. There are three types of DAT reagents available: a polyspecific reagent that contains both anit-IgG with anti-C3d and monospecific reagents that contain either anti-IgG or anti-C3d. For the DAT assay, the patient's RBCs are incubated with a DAT reagent following a wash step. A positive reaction with anti-IgG (presence of agglutination) indicates the presence of antibodies bound to the RBCs in the patient's circulation (Fig. 4.2).

If the patient has recently been transfused (last 3 months) then the positive DAT could represent alloantibodies to the donor RBCs in the patient's circulation and if there is no recent transfusion history then the positive DAT could represent autoantibodies. If the DAT is negative the positive AC could be attributed to antibodies to the enhancement media. In addition to these findings, a positive DAT could also be related to drugs or autoimmune disease that could be further elucidated by the patient's history. If the DAT is positive with the anti-IgG reagent then an elution should be performed.

Elution

Elution is a procedure that allows for the uncoupling (elution) of bound antibodies from sensitized RBCs. These now unbound (eluted) antibodies are referred to as the eluate. There are various techniques that can be used for the eluting of antibodies from sensitized RBCs such as heat elution, acid elution, or Lui freeze elution. Once the eluate is obtained it should be run against an extended antibody panel to determine the specificity of the antibodies.

Adsorption studies

Warm autoantibodies are characterized by the presence of a panreactive panel plus a positive AC/DAT with the eluate showing the same reactivity on the extended panel. Due to the panreactivity seen on the panel determining the presence of underlying

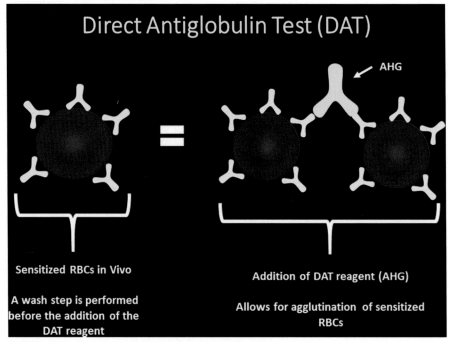

FIGURE 4.2 Principles of direct antiglobulin test (DAT).

alloantibodies is not possible. Adsorption is a technique that be used to remove the autoantibody from the patient's serum. If the patient has not been recently transfused (last 3 months) then the patient's own cells can be used for the absorption studies—autologous adsorption also referred to as an autoabsorption. However, if there is a recent transfusion history or if history is unknown then an allogeneic adsorption must be performed. The "absorb serum" can then be used to identify underlying alloantibodies by running the sample on the extended panel.

Phenotyping and Genotyping

Phenotyping is a serologic technique used to identify the antigens expressed on a patient's RBCs. In patients that have a history of multiple alloantibodies providing phenotype blood may be an acceptable option as that patient is known to be a responder due to formation of multiple alloantibodies as well as the difficulty that may be encountered when trying to determine the presence of other underlying alloantibodies. However, in patients that have been recently transfused, phenotyping should not be performed as the phenotype resulted may represent the transfused cells. In this case genotyping should be performed. Genotyping is a molecular method that can determine the RBC antigen profile, phenotype, by DNA testing obtained from white blood cells.

INTERPRETATION OF SEROLOGY FINDINGS

Antibodies that react at IS phase are more likely to be IgM, cold reacting, nonclinically significant antibodies; with the exception being ABO antibodies. Antibodies reacting at AHG phase are more likely to be IgG, warm reacting, clinically significant antibodies.

Common Reaction Patterns on Antibody Identification Panel

If the extended panel demonstrates selective cells that are reactive with the same strength then this likely represents a single alloantibody. If this pattern is seen in conjunction with a positive AC, this could represent an alloantibody in someone recently transfused and further testing with DAT/eluate should be performed with the eluate showing a similar pattern on the extended panel. If selective cells are reactive with variable strength this could represent a single alloantibody showing dosage or multiple alloantibodies. If panreactivity is seen with variable strength this could represent multiple alloantibodies. If panreactivity is seen with uniform strength and a negative autocontrol this could represent alloantibodies to high frequency antigens or antibodies to the preservatives in the reagent RBCs. If panreactivity is seen with uniform strength as well as a positive AC this may represent warm autoantibodies and DAT/eluate should be performed. Transfusion history should also be obtained as this will aid in determining if this is a warm autoantibody or possibly an alloantibody to a high frequency antigen in the setting of recent transfusion. If the patient has a positive panreactive panel with a positive AC but negative DAT this could represent antibodies to the enhancement media. A single reactive reagent cell could represent alloantibodies to low-frequency antigens.

KEY POINTS

- A collected blood sample for pretransfusion testing should include two unique identifiers for the patients (typically name and MRN) as well as the date of collection and phlebotomist collecting the sample. The date of collection and person collecting the specimen do not necessarily have to be mentioned on the label of the specimen as long as there is a process in place to identify this information.
- For blood bank testing, blood should be collected in either a pink top EDTA tube for plasma or a red top tube for serum.
- Patient samples that are processed through the blood bank must be retained for at least 7 days and should be stored at refrigerator temperatures.
- A new pretransfusion sample is required every 3 days when the patient has been transfused or pregnant in the last 3 months, or the history is unknown.
- The "type" includes determination of the patients ABO antigens and ABO antibodies (isohemagglutinins) and as well Rh status, specifically the D antigen status.

- The "screen" is a limited panel that contains the major clinically significant RBC antigens and is used to identify the presence or absence of alloantibodies. If positive, an extended panel must be performed to identity the specificity of the alloantibody/alloantibodies causing the positive screen.
- ABO antigens and ABO antibodies (isohemagglutinins) are the most clinically significant blood group.
- A patient's ABO antigens are determine by forward typing while isohemagglutinins are determine by the reverse typing.
- If an ABO discrepancy is present, universal blood components (type O RBCs and type AB plasma) should be issued until the cause of the ABO discrepancy is determined.
- The purpose of the crossmatch is to demonstrate compatibility of the selected RBC blood product with the patient's plasma (major crossmatch) or compatibility of the selected plasma blood product with the patient's RBCs (minor crossmatch).
- If the patient has no history of alloantibodies and the current pre-transfusion sample does not demonstrate alloantibodies, then ensuring ABO compatibility is the only requirement at the time of crossmatching. ABO compatibility can be demonstrated by either an immediate spin crossmatch or an electronic crossmatch.
- If the patient has a history of alloantibodies and/or the current pre-transfusion sample demonstrates alloantibodies, then an AHG crossmatch must be performed for RBC containing blood products.

REFERENCES

[1] Fung M, Grossman B, Hillyer C, Westhoff C. Technical manual, 18th ed. American association for Blood Banks (AABB); 2014.
[2] Bolton-Maggs PH, Wood EM, Wiersum-Osselton JC. Wrong blood in tube-potential for serious outcomes: can it be prevented? Br J Haematol 2014;168:3–13.
[3] Figueroa PI, Ziman A, Wheeler C, Gornbein J, et al. Nearly two decades using the check-type to prevent ABO incompatible transfusions: one institution's experience. Am J Clin Pathol 2006;126:422–6.
[4] Lumadue JA, Boyd JS, Ness PM. Adherence to a strict specimen-labeling policy decreases the incidence of erroneous blood grouping of blood bank specimens. Transfusion 1997;37:1169–72.
[5] Mijovic A. Alloimmunization to RhD antigen in RhD-incompatible haemopoietic cell transplants with non-myeloablative conditioning. Vox Sang 2002;83(4):358–62.
[6] Schwartz J, Padmanabhan A, Aqui N, Balogun RA, et al. Guidelines on the use of therapeutic Apheresis in clinical practice-evidence-based approach from the writing committee of the American Society for Apheresis: the seventh special issue. J Clin Apher 2016;31:149–338.

Red cell antigens and antibody

INTRODUCTION

Blood group antigens are organized into systems, collections, low incidence antigens, and high incidence antigens. There are 328 antigen specificities, 284 of which belong to one of the 30 blood group systems. A blood group system is defined as one or more antigens controlled at a single gene locus or closely linked homologous genes. The remainder of the chapter will focus on the major clinically significant blood group systems.

H BLOOD GROUP SYSTEM

The H blood group system, ISBT number (018)/symbol (H)/CD number 17, contains one high prevalent (H) antigen encoded by the *H* (*FUT1*) gene. The *H* gene (*FUT1 or FUT1*01*) is located on chromosome 19q 13.33 and consists for four exons distributed over 8 kbp of genomic deoxyribonucleic acid (gDNA). The primary gene product of the H gene is 2-α-fucosyltransferase, an enzyme that adds the sugar α-L-fucose on to the terminal galactose of type 2 carbohydrate precursor chains attached to protein or lipids on cells. Addition of this immunodominant sugar results in the formation of the H antigen on red blood cells (RBCs). The homologous gene (*Se* or *FUT2* or *FUT2*01*), located 35 kbp closer to the centromere on chromosome 19q 13.33, is responsible for adding the sugar α-L-fucose on to the terminal galactose of type 1 carbohydrate precursor chains attached to protein or lipids in secretions. Addition of this immunodominant sugar results in the formation of the H antigen in secretions or plasma. This is functionally referred to as being a secretor. The fucosylated glycans products of *FUT1* and *FUT2* are believed to serve as ligands in cell adhesions or as receptors for certain microorganisms. The H antigen is broadly distributed in tissues, platelets, RBCs, lymphocytes (in secretors), saliva, and all body fluids (in secretors) except cerebrospinal fluid (CSF). The H antigen has been shown to play a role in cell adhesion, normal hematopoietic differentiation, and in several malignancies [1].

ANTIGENIC DETERMINANTS

The reference allele of the *FUT1* gene (FUT1*01) *encodes 2-α-* fucosyltransferase which adds a fructose via α1-2 linkage to terminal galactose of type 2 chains on RBCs and other cells. Numerous alleles of the *FUT1* gene exist and alternation of the reference gene results in the null (Bombay, O_h) phenotype and in rare instances H+^{w} (Para-Bombay). The reference allele of the *FUT2* gene (FUT2*01) *also encodes 2-α-* fucosyltransferase which adds a fructose via α1-2 linkage to terminal galactose of type 1 chains in secretions. Numerous alleles of the *FUT2* gene exist and alternation of the reference gene results in the Para-Bombay phenotype or H+^{w} (weak expression of the H antigen in secretions). Some cross-reactivity may occur but generally speaking the two enzymes prefer the precursor chain indicated above. The H antigen is a high frequency antigen. According to the literature, it has a prevalence of 99.9% in all populations on RBCs. It is weakly expressed on cord cells. And since the H antigen expression is partially masked if a functional A or B allele is inherited, the amount of H expression is as follows in decreasing order $O > A_2 > B > A_2B > A_1 > A_1B > H+^w$. The H antigen is resistant to all proteolytic enzymes [1–3].

PHENOTYPE

As mentioned earlier, alternations of the reference genes *FUT1* and *FUT2* alters phenotypic expression of the H antigen. The three main phenotypes observed include secretor, nonsecretor, and the H-deficient phenotypes (Bombay and Para-Bombay). The characteristics of the main phenotypic expressions are summarized in Table 5.1. Individuals belonging to the nonsecretor phenotype have at least one function H gene (*FUT1*) and do not possess any function secretor genes (*FUT2*). As a result, individuals with this phenotypic expression have H antigen present on the surface of RBCs and do not have any H antigens in secretions. Individuals with secretor phenotype have the H antigen present both on the surface of RBCs and in secretions. H-deficient phenotypes include Bombay and Para-Bombay, representing very rare blood groups. The Bombay phenotype, first discovered by Bhende in 1952 in Bombay, India, is

Table 5.1 The Characteristics of the Main Phenotypic Expressions of H Antigen

Common Phenotypes	H Antigen on RBCs	H Antigen in Secretions	Genotype	Serum Antibodies
Secretor	Yes	Yes	HH or Hh; SeSe or Sese	Anti-HI
Nonsecretor	Yes	No	HH or Hh; sese	Anti-HI
Bombay	No	No	Hh; sese	Anti-H
Para-Bombay	No or Weak	Yes, Rare cases No	Hh or (H)*; SeSe or Sese or sese	Anti-H or Anti-HI
LADII	No	No	Any genotype	Anti-H

*Altered H gene with weak expression on RBC.

due to inheritance of two mutated *FUT1* and *FUT2* genes, genomically referred to as *hhsese*. Numerous silencing mutations of the *FUT1*01* and *FUT2*01* exist. According to genetic studies, these mutations typically occur primarily in exon 4 in *FUT1*01* and in exon 2 in *FUT2*01*. Inheritance of either homozygous or compound heterozygous mutations in *FUT1*01* and *FUT2*01* leads to failure to encode the corresponding functional 2-α-fucosyltransferase. As a result, these individuals lack ABH antigens on RBCS and in secretions. Additionally, they naturally form anti-H and can only be transfused with blood from another Bombay individual. Individuals with mutations in the GDP-fucose transporter gene (*SLC35C1*) also present with a Bombay-like phenotype (H-) and lack A, B, Le^a, and Le^b. This is due to the GDP-fucose transporter being ineffective resulting in no fucosylation despite normal *FUT1*, *FUT2*, and *FUT3* genes. Changes in the *SLC35C1* gene have also been associated with white blood cells (WBCs) that lack CD15/sialyl-Le^x. These changes result in a rare leukocyte adhesion deficiency (LADII or CDGII). Individuals with LADII present with mental retardation, severe recurrent infections with a high WBC count, and form anti-H. Para-Bombay (H^{+W}) individuals are classically described as those who lack RBC ABH antigens on RBCs but possess them in sections. In other words, these individuals have an inactive *FUT1* and a functional *FUT2* gene but some Para-Bombay individuals have an altered but weakly active *FUT1* and an active or inactive *FUT2*. Thus, these individuals have very few ABH antigens on RBCs and may or may not possess them in secretions. Depending on the genetic phenomena, these individuals will either procedure anti-H or anti-HI alloantibodies [1,3,4]. The characteristics of the main phenotypic expressions of H antigen are summarized in Table 5.1. Major aspects of H blood group system are summarized in Table 5.2. Various features of H blood group are also presented in Fig. 5.1.

ABO BLOOD GROUP SYSTEM

The ABO blood group system, ISBT number (001)/symbol (ABO) was first discovered in the early 1900s by Karl Landsteiner. Using a rudimentary form of forward (antigen detection) and reverse (antibody detection) typing, he mixed RBCs and serum collected from his colleagues and based on the pattern of agglutination, originally proposed the ABO blood group system. Currently, the ABO system contains four polymorphic (A, B, AB, A1) antigens. These antigenic determinants are encoded by the *ABO* gene. Originally the ABO system contained five antigenic determinants, but the ABO5 antigen is made obsolete after the H antigen was removed to form the H system in 1990 [1].

The ABO gene is located on chromosome 9q 34.2 and consists for seven exons distributed over 19.5 kbp of gDNA. Over 200 alleles have been described at the ABO locus with three common A, B, and O alleles observed. The primary gene product of the ABO gene are not the A and B antigens but glycosyltransferases that covert that H substance (generated by *FUT1*01* and *FUT2*01*) to blood group A and B. The addition of *N*-acetyl-galactosamine added in $\alpha1 \rightarrow 3$ linkage to the subterminal galactose

Table 5.2 Major Features of H Blood Group System: ISBT Number (018)/ Symbol (H)

Antigens	1: H
Gene (FUT1)	Located on chromosome 19q 13.33
	Consists for four exons distributed over 8 kbp of gDNA
	Primary gene product of the H gene is 2-α-fucosyltransferase, an enzyme that adds the sugar α-L-fucose on to the terminal galactose of type two carbohydrate precursor chains attached to protein or lipids on cells
	The homologous gene (Se or FUT2 or FUT2*01), located 35 kbp closer to the centromere on chromosome 19q 13.33, adds the sugar α-L-fucose on to the terminal galactose of type 1 carbohydrate precursor chains attached to protein or lipids in secretions
Protein	Copies of the H antigen protein found on the surface of RBCs may vary.
	The fucosylated glycans products of FUT1 and FUT2 are believed to serve as ligands in cell adhesions or as receptors for certain microorganisms and been shown to play a role in cell adhesion, normal hematopoietic differentiation, and in several malignancies
	The H antigen is broadly distributed in tissues and platelets. RBCs, lymphocytes (in secretors), saliva, and all body fluids (in secretors) except CSF
Antigenic Frequencies	The H antigen expression is partially masked if a functional A or B allele is inherited, the amount of H expression is as follows in decreasing order O $>A_2 > B > A_2B > A_1 >A_1B > H+^w$
Phenotypes	See Table 5.1
Antibodies	IgM
	Some complement binding
	Immediate and delayed hemolytic transfusion reactions

CSF, Cerebrospinal fluid; gDNA, genomic deoxyribonucleic acid.

of the H antigen results in the formation of the A antigen, whereas addition of galactose added in $\alpha1\rightarrow3$ linkage to the subterminal galactose of the H antigen results in the formation of the B antigen. Blood group O results when a mutation of the ABO gene leads to no further modification of the H antigen. ABO antigens are located on RBCs, endothelial cells, kidney, heart, pancreas, platelets, and lungs. In secretors, they are also found in saliva and all body fluids except the CSF. The function of A and B glycosyltransferase is unclear but some investigators believe that the antigenic variability has a function in herd immunity, while others postulate they play a role in embryogenesis, cell–cell interaction in carcinogenesis, and modulation of sialic acid recognition. Interestingly, A and B antigens can show weakened expression in pregnancy, leukemia, myelodysplastic syndrome, and any disease that induces stress hematopoiesis. Complete loss of A and B antigens have been described various solid

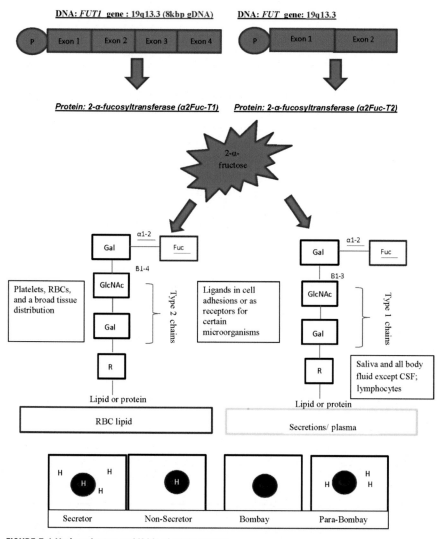

FIGURE 5.1 Various features of H blood group system.

tumors. Moreover, conversion of the A antigen to a B-like antigen, so called acquired B, is the consequence of microbial infection and deacetylation of the terminal Gal-NAc of the A antigen.

ANTIGENIC DETERMINANTS

The two main reference alleles of the ABO gene (*ABO*A1.01* and *ABO* B.01*) encode the glycosyltransferases that covert that H substance to blood group A and B. The addition of *N*-acetyl-galactosamine added in $\alpha1 \rightarrow 3$ linkage to the subterminal galactose of the H antigen results in the formation of the A antigen, whereas addition of galactose in $\alpha1 \rightarrow 3$ linkage to the subterminal galactose of the H antigen results in the formation of the B antigen. The A and B alleles are autosomal codominant and differ by seven nucleotides and four amino acids.

Numerous alleles of the *ABO*A1.01* and *ABO* B.01* genes exist. Alternation of the reference gene results in the presence of numerous weak A and B subgroups, cis-AB, B (A), A (B), and null (group O) phenotypes.

The A antigen is present in high frequency among Caucasians. The prevalence is 43% in Caucasians, 27% in Blacks, and 27% in Asians. The two main subgroups are A1 and A2. Approximately 80% of blood group A individuals are A1 with nearly 20% representing A2. A1 individuals are characterized by having 1 million A antigenic epitopes per RBC, whereas A2 individuals poses 1/5 the number of antigenic sites (2.2×10^5). This quantitative difference in A antigen expression is due to the A1 enzyme being more active (5–10 times) then the A2 enzyme. It also results in qualitative differences. Therefore, A1 can be distinguished from A2 by the lectin *Dolichos biflorus*, which will agglutinate A1 but not A2 cells. Additionally, since the A2 enzyme is less efficient in the conversion of the H antigen, RBCs of A2 individuals will show increased reactivity with anti-H lectin *Ulex europaeus*. The B antigen has a prevalence of 9% in Caucasians, 20% in Blacks, and 27% in Asians. Alternations of the *ABO* B.01* allele results in B antigen subgroups but the rarity of this occurrence is not usually of clinical significance. ABO antigens can be detected on RBCs as early as 5–6 weeks of gestation and reach adult levels of expression by 2–4 years. They are resistant to all proteolytic enzymes.

PHENOTYPE

As mentioned earlier, alternations of the reference alleles also alters phenotypic expression of the ABO antigens. The four main phenotypes observed include O, A, B, and AB. Although numerous subgroups of the A antigen exists but only subgroups A1 and A2 are observed frequently. Rarer phenotypes such as B(A) and acquired B phenotypes also exist. Frequencies of main phenotype of ABO antigens are summarized in Table 5.3.

Since the *ABO*A1.01* and *ABO* B.01* alleles are codominantly expressed and the O antigen represent an amorphic allele encoding a nonfunctional enzyme, genotypic expression of the A and B phenotypes can vary. The A phenotype can be acquired by inheritance of either two functional *ABO*A1.01* alleles (AA) or one functional *ABO*A1.01* allele and one amorphic nonfunctional allele (AO). Numerous mutations in *ABO*A1.01* allele can result in formation of the A2 phenotypes, the most common involving 1061del C polymorphism. Likewise, the B phenotype can be acquired by inheritance of either two functional *ABO* B.01* alleles (BB) or one functional *ABO* B.01* allele and one amorphic nonfunctional allele (BO), whereas the O blood group phenotype is due to homozygous or compound heterozygous amorphic ABO (OO) alleles. The AB phenotype is due to inheritance of one functional *ABO*A1.01* alleles

Table 5.3 The Characteristics of the Main Phenotypic Expressions of ABO Antigen

Phenotype	Caucasians (%)	Blacks (%)
O	43	49
B	9	20
A_1	33	19
A_2	10	8
A_1B	3	3
A_2B	1	1

(A) and one functional *ABO*B.01 allele*. Since the ABO system is characterized by the presence of naturally occurring alloantibodies, each phenotype develops alloantibodies against the missing A and B antigens. This is believed to be due to exposure to immunizing sources such as gut and environmental bacteria that possess ABO-like structures on their lipopolysaccharide coats.

B(A) phenotype is characterized by having weak expression of A antigen on group B cells. Polymorphisms in critical amino acids result in increased capacity of the B glycosyltransferase to use UDP-*N*-acetylgalactosamine, in addition to UDP-galactose resulting in A antigen synthesis. The acquired B phenotype is transient phenomena in where group A individual demonstrates weak B expression. This is due to deacetylation of the A antigen yielding a B-like sugar and is typically seen in the setting of gastrointestinal bacterial infections [1–4]. Major characteristics of ABO blood group system are summarized in Table 5.4. Various features of ABO blood group are also presented in Fig. 5.2.

LEWIS BLOOD GROUP SYSTEM

The Lewis blood group system, ISBT number (007)/symbol (LE) was first discovered by Mourant in 1946. It currently contains six antigens which include five polymorphic (Le^a, Le^b, Le^{bH}, ALe^b, and BLe^b) antigens and one high prevalence (Le^{ab}) antigen. These antigenic determinants are encoded the *LE* (*FUT3*) gene [1].

The *LE* gene (*FUT3*) is located on chromosome 19q 13.3 and consists for three exons distributed over 8 kbp of gDNA. The primary gene product of the *LE* gene is α (1, 3/4) fucosyltransferase, an enzyme that adds the sugar fucose in an α1→4 linkage to penultimate *N*-acetyl-glucosamine on type 1 carbohydrate precursor chains attached to protein or lipids on cells. Depending on the sequential actions of the *FUT2* and *FUT3* transferases, either the Le^a or Le^b antigen will be formed. If the *FUT2* gene is functionally active first, the Le^b antigen is generated. If the *FUT3* trans-

Table 5.4 Major Features of ABO Blood Group System: ISBT Number (001)/ Symbol (ABO)

Antigens	4: A, B, AB, A1
	The antigen determinants built upon the H antigen
Gene (ABO)	Located on chromosome: 9q34.1q34.2
	Consists for seven exons distributed over 19.5 kbp of gDNA
Protein	ABO antigens are located on RBCs, endothelial cells, kidney, heart, pancreas, platelets, and lungs. In secretors, they are also found in saliva and all body fluids except the CSF. The function of A and B glycosyltransferase is unclear but some believe the antigenic variability has a function in herd immunity,while others postulate they play a role in embryogenesis, cell–cell interaction in carcino-genesis, and modulation of sialic acid recognition
Phenotypes	See Table 5.3
Antibodies	IgM
	Bind complement
	Severe hemolytic transfusion reaction
	Mild hemolytic disease of the fetus and newborn (HDFN)

CSF, *Cerebrospinal fluid;* gDNA, *genomic deoxyribonucleic acid.*

ferase functions prior to the action of *FUT2*, the addition of subterminal fructose sterically inhibits the binding of the secretor enzyme and Le^a is formed. A functionally inactive *FUT2* gene in the setting of a functionally active *FUT3 gene* also results in formation of Le^a. The ALe^b and BLe^b antigens are the result of the sequential actions of the *FUT2*, *FUT3*, and ABO.

The Le antigens are found in a broad distribution. They are found on lymphocytes, platelets, endothelium, and on the epithelial of the kidney, genitourinary, and gastrointestinal tract. These antigens are also found in saliva and all body fluids except the CSF. Lewis antigens are not intrinsic to the RBC membrane but are instead passively absorbed from of the glycoproteins in the serum. Thus, expansion of plasma during pregnancy has been shown to transiently result in a Le (a−b−) phenotype. Lewis antigens are also lost from RBCs in the following disease states: infectious mononucleosis complicated with hemolysis, severe alcoholic cirrhosis, and alcoholic pancreatitis. The Lewis antigens are speculated to serve as ligands for E-selectins and have been implicated as the receptor for several infectious pathogens. The Le^b antigen is the receptor for *Helicobacter pylori* and Norwalk. Additionally, increases susceptibility to *Candida* and other uropathogenic *Escherichia coli* infections has been seen in individuals with the Le (a−b−) phenotype.

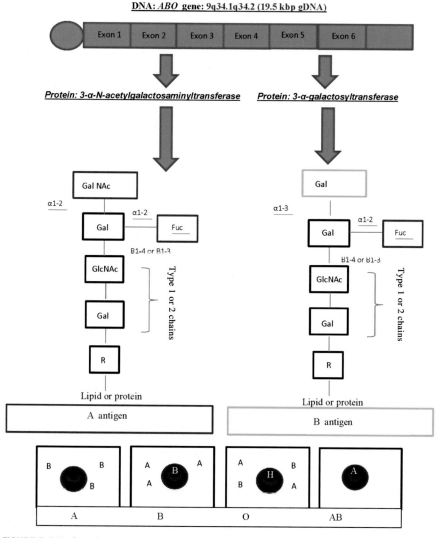

FIGURE 5.2 Various features of ABO blood group system.

ANTIGENIC DETERMINANTS

The two main antigenic determinants, Le^a or Le^b, differ structurally by only one fucose. Le^a antigen was identified in 1946 and has a prevalence of 22% in Caucasians and 23% in Blacks. The Le^B antigen was identified in 1948 and has a prevalence of 72% in Caucasians and 55% in Blacks. Both antigens are not present on cord cells and resistant to all proteolytic enzyme, particularly ficin, trypsin, and chymotrypsin.

PHENOTYPE

The three main phenotypes observed include Le(a+b−), Le(a−b+), and Le(a−b−). A fourth phenotype, Le(a+b+) is rarely seen. Expression of these phenotypic profiles are based on the functionally of the *FUT2* and *FUT3* genes. The phenotypic frequencies of Lewis antigens are summarized in Table 5.5.

The Le (a+b−) phenotype is due to inheritance of at least one functional Lewis gene and a nonfunctional secretor allele. Thus, these individuals also lack the ability to synthesize type I ABH antigens. The Le(a−b+) phenotype is due to the inheritance both a function *FUT 2* and *FUT3* gene. These individuals have the ability to make Le^a, Le^b, and type 1ABH chains. Due to the efficiency of the *FUT2* gene, most of the precursor type 1 chain is converted to Le^b thus these individuals' RBCs are negative for Le^a. Additionally, despite typing negative for Le^a, these individuals do not make anti-Le^a. The Le(a+ b+) phenotype is due to weak secretor gene. This can be transient, as seen in children, who initially type Le (a−b−) due to developmental delays in *FUT2* activity or it can be due to a mutation in the *FUT2* gene with decreased efficiency. This phenotypic expression is usually seen in the Japanese population at a rate of 16%. The Le(a−b−) phenotype may result from more than 30 different null alleles at the *FUT3* locus. These individuals lack the functional ability to make Le^a and Le^b but can still make Type 1ABH if they have a function *FUT2* gene. Mutations in the *FUT6* and GDP-fucose transporter also result in the null Lewis phenotype. Due to the lack of Lewis antigen, these individuals can form anti-Le^a, -Le^b, and -Le^{a+b} [1–4]. Major aspects of Lewis blood group system are summarized in Table 5.6. Various featured of Lewis blood group system are also presented in Fig. 5.3.

I BLOOD GROUP SYSTEM

The I blood group system, ISBT number (027)/symbol (I) was discovered in 1956 and placed in its own system in 2002. It was named I to emphasize its high degree of individuality. Currently, the I system contains one high prevalence (I). This antigenic determinant is encoded by the *I (GCNT2, IGnT)* gene. The Ii collection, ISBT number (207)/symbol (I), currently contains one high prevalence antigen (i), named in 1960 for its reciprocal relationship to I. The genetic basis for this antigen is unknown. Previously, both I and i belonged to the Ii collection but I was promoted

Table 5.5 The Phenotypic Frequencies of Lewis Antigens

Phenotype	Caucasians	Blacks
Le(a+b−)	22%	23%
Le(a−b+)	72%	55%
Le(a−b−)	6%	22%
Le(a+b+)	Rare	Rare

Table 5.6 Major Features of Lewis Blood Group System: ISBT Number (006)/Symbol (Le)

Antigens	Six: five polymorphic (Le^a, Le^b, Le^{bH}, ALe^b, and BLe^b) and one high prevalence (Le^{ab}) antigens
	Synthesized on type 1 precursor chains
	Sequential actions of the *FUT2, FUT3,* and ABO
Gene (FUT3)	Located on 19q 13.3
	Consists for three exons distributed over 8 kbp of gDNA. The primary gene is α (1,3/4) fucosyltransferase, an enzyme that adds the sugar fucose in an α1 →4 linkage to penultimate *N*-acetyl-glucosamine on type 1 carbohydrate precursor chains attached to protein or lipids on cells
Protein	Located on lymphocytes, platelets, endothelium and on the epithelial of the kidney, genitourinary and gastrointestinal tract. They are also found in saliva and all body fluids except the CSF. Lewis antigens are not intrinsic to the RBC membrane but are instead passively absorbed from of the glycoproteins in the serum
	Lewis antigens are also lost from RBCs in the following disease states: pregnancy, infectious mononucleosis complicated with hemolysis, severe alcoholic cirrhosis, and alcoholic pancreatitis.
	Speculated to serve as ligands for E-selectins. And have been implicated as the receptor for several infectious pathgens. The Le^b antigen is the receptor for *Helicobacter pylori* and Norwalk. Additionally, increases susceptibility to *Candida* and other uropathogenic *Escherichia coli* infections has been seen in individuals with the Le (a−b−) phenotype
Antigenic Frequencies	Le^a: 22% Caucasians, 23% Blacks
	Jk^b: 72% Caucasians, 55% Blacks
Phenotypes	See Table 5.5
Antibodies	IgM
	Naturally occurring
	Seen transiently in pregnancy
	Le(a+b−) rarely makes anti-Le^b, Le(a−b+) do not make anti-Le^a, Le(a−b−) make anti-Le^b, -Le^a, and -Le^{a+b}

CSF, *Cerebrospinal fluid;* gDNA, *genomic deoxyribonucleic acid.*

to its own blood group system, when gene encoding the branching transferase (*GCNT2*) was cloned [1].

The *GCNT2,* gene is located on chromosome 6p24.2 and consists for five exons, three of which are tissue specific exons (exon 1A, 1B, 1C), distributed over 100 kbp of gDNA. The primary gene product of the *GCNT2* gene is 6-*β-N*-acetylgalactosaminyltransferase. This enzyme converts the liner i antigen into a branched I antigen. As mentioned above the genetic background of the i antigen is unknown. It is derived

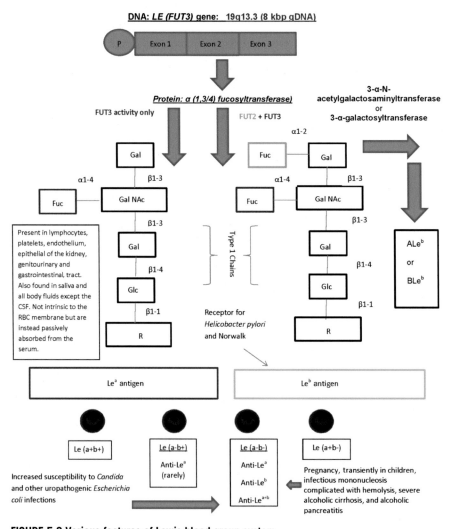

FIGURE 5.3 Various features of Lewis blood group system.

from Type 2 chains by the addition of at least 2 [Gal(β1-4)GlcNAc (β1-3)] segments. Further modification of the i antigen by 6-β-N-acetylgalactosaminyltransferase and UDP galactose generates the I antigen. Structurally, the I antigen differs from the i antigen due to addition of N-acetylglucosamine in β1-6 linkage and the addition of galactose. Further modification can occur and result in the expression of ABH and sialic acid. The I and i antigens are ubiquitously on all cell membranes. The exact copies of the I antigen on adult RBCs varies. Studies have shown anywhere from $0.3\text{--}5.0 \times 10^5$ copies are expressed. The number of i antigen also varies, with $0.2\text{--}0.65 \times 10^5$ on cord cells and $0.3\text{--}0.7 \times 10^5$ on i_{adult} cells. In addition to being present on RBCs and having a wide tissue distribution, I and i antigens are also present on leukocytes, platelet and in plasma, saliva, human milk, amniotic fluid, urine, and ovarian cyst fluid. However, the function of the I and i antigen is unknown, but both antigens appear to serve as substrate for the synthesis of ABO, Lewis X, and other type 2 chain antigens. This results in the formation of compound antigens such as IA, IB, IAB, iH, IH, iP1, IP1, iHLe^b, and ILe^{bH}.

ANTIGENIC DETERMINANTS

As mentioned previously the i antigen is a repeating linear type 2 chain structure. While, the I antigen is a branched chain structure derived from the i antigen, neonatal RBCs predominantly express the i antigen. However, due to phosphorylation of key transcription factors which active *GCNT2,* with age conversion of the i antigen occurs. This results in an increasing amount of the I antigen and a decreasing amount of the i antigen. The I antigen has an occurrence of >99% in adult, weakly expressed on cord cells and resistant to enzyme treatment. In contrast the i antigen strongly expressed on cord cells, trace amounts are typically found on adults, and is resistant to enzyme treatment.

PHENOTYPE

There are two recognized phenotypic expressions of the I antigen: I (I+) and i (I−). Typically, children develop the adult I+ phenotype by the age of 2 years of age. An increased amount of i antigen can occur as a sign of stressed erythropoiesis secondary to a chronic hemolytic disorder [leukemia, Tk polyagglutination, thalassemia, sickle cell disease, Diamond Blackfan anemia, myeloblastic erythropoiesis, sideroblastic erythropoiesis, and HEMPAS (enzyme defect in *N*-glycosylation resulting in chronic hemolysis, splenomegaly, and erythroid multinuclearity)] and is reflective rapid transit through ER and Golgi. It may also be due to genetic mutations in the *GCNT2* gene (i$_{adult}$ or null phenotype). There are seven different point mutations that lead to the i$_{adult}$ (null phenotype). Two variants exist of the i$_{adult}$ phenotype exist and differentiate by the presence of cataracts. i$_{adult}$ with cataracts form is due either a gene deletion or mutation in exons 2 and 3, seen in those with Asian ancestry, and results in the loss of I antigen synthesis in all tissues, whereas the i$_{adult}$ without cataracts form is due to a mutation in exon 1c. As a result, I antigen is missing on RBCs but present in all other tissues. It is typically seen in Caucasians. Phenotypes of I antigen are summarized in Table 5.7.

In the setting of the adult i phenotype, an alloanti-I can develop. Accordingly to the literature, this is rare and has not been implicated in HDFN or transfusion reactions. More commonly, autoanti-I is seen in the serum. This auto-antibody is cold, naturally occurring, and IgM in nature. It is usually clinically insignificant, has titers <64 and is seen in the serum of many healthy individuals. As expected autoanti-I reacts more strongly with adult cells then cord cells. On occasion, autoanti-I can be pathologic. Specifically, autoanti-I has been associated with cold hemagglutinin

Table 5.7 Phenotypes of I Antigen

Phenotype	Frequency	Antigen	Disease Association
Adult	>99% in adults	I (strong) i (weak)	
Cord	All	I (weak) i (strong)	
i adult	Rare	I (weak) i (strong)	i_{adult} with cataracts form is due either a gene deletion or mutation in exons 2 and 3, seen in those with Asian ancestry, and results in the loss of I antigen synthesis in all tissues. i_{adult} without cataracts form is due to a mutation in exon 1c. This results in I antigen missing on RBCs but being present in all other tissues. It is typically seen in Caucasians

disease (CHAD) and pneumonia due to *Mycoplasma pneumoniae*. Autoanti-i is not commonly observed in the sera of healthy individuals. But when it is present, like autoanti-I, it is reactive at 4°C, IgM in nature, and reacts more strongly with cord cells. Autoanti-i can also be pathogenic and associated with infectious mononucleosis. Horse RBCs have strong expression of i antigen and can be used as a diagnostic tool for infectious mononucleosis [1–4]. Major aspects of I blood group system are summarized in Table 5.8.

P1PK AND GLOBOSIDE BLOOD GROUP SYSTEMS

The P1PK blood group system, ISBT number (003)/symbol (P1PK) was first discovered in the early 1927 by Landsteiner and Levine in a series of experiments that also lead to the discovery of the M and N antigens. Currently, the P1PK system contains one polymorphic (P1), one high prevalence (P^k), and one low prevalence (NOR) antigen. These antigenic determinants are encoded by the *A4GALT* gene. Originally, the P1PK system contained the P antigenic determinant but in 2002 molecular studies necessitated it to be moved to the Globoside (GLOB) collection. The GLOB collection, ISBT number (028)/symbol, currently contains one high prevalence antigen (P) and the PX2 antigen, which are encoded by the *GLOB* gene. The LKE antigen which is now included in the 209 collection system was previously included in the P1PK and GLOB systems. These antigenic determinants are related glycosphingolipids [1].

The *A4GALT* gene is located on chromosome 22q13.2 and consists for four exons distributed over 26.6 kbp of gDNA. The primary gene product of the *A4GALT* gene is 4-α-galactosyltransferase. This enzyme transfers a galactose to the terminal sugar of a paragloboside or a lactosylceramide to generate the P1 antigen or P^k, respectively.

Table 5.8 Major Features of I Blood Group System: ISBT Number (027)/ Symbol (I)

Antigen	1 antigen: I (i belongs to collection Ii group)
Gene (GCNT2)	located on chromosome 6p24.2
	Consists for five exons, three of which are tissue specific exons (exon 1A, 1B, 1C), distributed over 100 kbp of gDNA
	Gives rise to three transcripts: IGnTa, IGnTB, and IGnTC
	Gene for i is unknown
Protein	6-β-N-acetylgalactosaminyltransferase
	Adds N-acetylglucosamine in β1-6 linkage + action UDP galactose → I from i
	i is derived from Type 2 chains by the addition of at least two [Gal(β1-4)GlcNAc (β1-3)] segments
Antigenic Frequencies And phenotypes	See Table 5.7
Antibodies	Adult i phenotype → alloanti-I (rare, not implicated in HDFN or transfusion reactions)
	Autoanti-I: common, cold, naturally occurring, IgM, clinically insignificant antibody w/titers <64
	Pathologic autoanti-I: associated with CHAD and *Mycoplasma pneumoniae*.
	Autoanti-i: uncommon: cold, naturally occurring, IgM, clinically insignificant antibody w/titers <64
	Pathologic autoanti-i: associated with infectious mononucleosis

gDNA, *Genomic deoxyribonucleic acid.*

The *GLOB* gene is located on chromosome 3q26.1 and consists of at least five exons distributed over approximately 19 kbp of gDNA. The primary gene product of the *GLOB* gene is 3-β-N-acetylgalactosaminyltransferase. This enzyme acts in sequential action and transfers an N-acetylgalactosamine to a lactosylceramide (P^k antigen) to form GLOB (P antigen). From the P antigen, Luke (LKE), type 4 chain ABH antigen, and NOR are generated. P^k and P are found on RBCs, lymphocytes, granulocytes, monocytes, endothelium, placenta (trophoblasts and interstitial cells) fibroblasts, kidney, lung, heart, and synovium cells. However in the human body, P1 has only expressed on RBCs. P1-like substance in bird excrement, pigeon egg whites, and hydatid cyst fluid has been found. P^k, P, and LKE are high incidence antigens and are expressed all RBCs except those rare null phenotypes (p, P_1^K, P_2^K). However, the number of P1 and P^k antigens on the surface of RBCs is highly variable. Approximately, 1,500,000 copies of the P antigen are present on the surface of RBCs accounting for approximately 6% of total RBC lipid content. In general, these glycosphingolipid antigens functions as receptors for various organisms and play a role on the pathogenesis of disease. P1, P^k, P, and LKE antigens all serve as receptors for P-fimbriated uropathogenic *E. coli*. Thus, the presence of these antigens may influence susceptibility to urinary tract infection (UTI). Both P1 and P^k are receptors for *Streptococcus suis*, the causative agent of bacterial meningitis. P^k is also the receptor for Shiga toxins, the causative agent for shigella dysentery, and *E. coli*-associated hemolytic uremic syndrome. Additionally, P^k has been shown to be physiologic receptor involved in signal modulation of α-interferon receptor and CXCR4 (HIV co-receptors) thus, it may provide for protection against HIV-1 infection; while the P antigen has been implicated as the receptor for the Parvovirus B19 virus which is the etiologic agent of erythema infectiosum (fifth disease).

ANTIGENIC DETERMINANTS

The P1 antigen has a prevalence of 79% in Caucasians and 94% in Blacks. It is found on fetal RBCs as early as 12 weeks but weakens with gestational age. Thus, this antigen is poorly expressed at birth. Due to quantitative variation, the antigenic expression of the P1 antigen in adults varies in strength from one individual to another. Decreased P1 expression is also seen individuals who inherit the gene *In* (*Lu*) and with storage of blood. P1 is resistant to enzyme treatment. The P^k antigen is expressed on most RBCs. Like P1, it is found on cord cells and resistant to enzyme treatment. The P antigen has a prevalence of >99.9%, is found on cord cells, and is resistant to enzyme treatment.

PHENOTYPE

The combined expression of P1, P^k, and P antigens results in the five phenotypic expressions of P. These include P_1, P_2, P_1^k, P_2^k, and p. Phenotypes of P1PK are summarized in Table 5.9. P_1 is the most common phenotypic expression with a frequency of around 80%. This phenotype expresses all three antigens (P, P1, P^k) and has no propensity to form alloantibodies. P_2 is the second most common of the phenotypic

Table 5.9 Phenotypes of P1PK

Phenotype	Frequency	Antigen	Antibody	Antibody Association
P_1	80%	P, P1, and P^k	None	None
P_2	20%	P and P^k	Anti-P1	Hydatid, fascioliasis, and bird handlers
P_1^k	Rare	P1 and P^k	Anti-P	PCH and spontaneous abortion
P_2^k	Rare	P^k	Anti-P and Anti-P1	Spontaneous abortion (anti-P)
p	Rare	none	Anti-PP1P^k	Spontaneous abortion

PCH, *Paroxysmal cold hemoglobinuria.*

expression with an incidence of nearly 20%. These individuals express only the P and P^k antigens. Thus, they can generate an anti-P1 alloantibody. This is seen in approximately one-quarter to two-thirds of individuals with the P_2 phenotype. Anti-P1 is a naturally occurring IgM alloantibody frequently seen in the serum of patients with hydatid cyst disease, fascioliasis (liver fluke), and bird handlers. Antibody reactivity has been shown to be neutralized by hydatid cyst fluid and pigeon egg whites and enhanced by enzyme-treated red cells. It is considered clinically insignificant as it has not been shown to cause HDFN. However, rarely, when found to be reactive at 37°C, it has been implicated in hemolytic transfusion reactions necessitating crossmatch compatible units. The remaining three autosomal recessive phenotypes are rarely seen. Inactivating mutations in the 4-α-galactosyltransferase result in the p phenotype which is characterized by lack of P1, P, and P^k antigen. Those with the p phenotype can form an anti-PP1P^k (aka anti-Tj^a) which is a naturally occurring, combined antibody with antigenic specificity toward P, P1, and P^k, and it has been associated with early spontaneous due to the presence of P^k and P antigens on the placenta, HDFN, and hemolytic transfusion reactions. Inactivating mutations in the 3-β-N-acetylgalactosaminyltransferase result in the P_1^k phenotype which is characterized by lack of the P antigen. While, the P_2^k phenotype (P^k antigen only) is the result of a SNP that induces an alternative start codon. Both of these phenotypic profiles have the tendency to generate anti-P, which is naturally occurring, predominantly IgM (with some IgG) alloantibody. Like anti-PP1P^k, anti-P is clinically significant as it has been associated with hemolytic transfusion reactions, HDFN, and early spontaneous abortions. Autoantibody with anti-P specificity has also been reported in the setting of paroxysmal cold hemoglobinuria (PCH). PCH is typically a disease of children induced by a viral illness that presents as the sudden presence of hemoglobin in the urine typically after exposure to cold temperatures. This is due to the biphasic binding of the autoanti-P IgG antibody (it has a tendency to bind to red cells at colder temperatures and cause hemolysis at body temperature). This characteristic is demonstrated in the Donath-Landsteiner test [1–4]. Phenotypes of P1PK antigens are summarized in Table 5.9. Various aspects of P1PK and GLOB Blood group systems are summarized in Table 5.10.

RH SYSTEM

The Rh blood group system, ISBT number (004)/symbol (RH)/CD number (CD240D/CD240CE) is complex and contains many antigens that are highly immunogenic. It was first discovered in 1939 but confirmed in 1940 in a series of experiments performed by Landsteiner and Weiner in which they evaluated the immunologic response of rabbits injected with RBCs collected from rhesus monkey. It currently contains 52 antigens which include 14 polymorphic (D, C, E, e, F, Ce, G hr^s, C^G, RH26 (c-like), cE, hr^B, and Rh41), 26 low prevalence (C^w, C^x, V$^\wedge$, E^w, VS$^\wedge$, CE, D^w, hr^h, Go^a, Rh32, Rh33, Rh35, Be^a, Evans, Tar, Crawford, Riv, JAL, STEM, FPTT, BARC, JAHK, DAK$^\wedge$, LOCR, and CENR), and 12 high prevalence (Hr$_0$, Hr, Rh29, H^b, Rh39, Nou, Sec, Dav, MAR, CEST, CELO, and CEAG) antigens. Rh-associated glycoprotein

Table 5.10 Major Features of P1PK and GLOB Blood Group Systems: ISBT (003/028)/Symbol (P1PK/GLOB)

P1PK: 3: P1, P^k, and NOR
GLOB: 2: P and P2X

Antigen	Gene	Chromosome	
P1/P^k/NOR	A4GALT	22q13.2	Four exons distributed over 26.6 kbp of gDNA
P	*GLOB*	3q26.1	Five exons distributed over approximately19 kbp gDNA

4-α-galactosyltransferase transfers a galactose to the terminal sugar of a paragloboside or a lactosylceramide to generate the P1 antigen or P^k, respectively. 3-β-N-acetylgalactosaminyltransferase transfers an N-acetylagalactosamine to a lactosylceramide (P^k antigen) to form globoside (P antigen). From the P antigen, Luke (LKE), type 4 chain ABH antigen, and NOR are generated.

P1: 79% Caucasians and 94% Blacks
P^k: Most RBCS

P: >99.9% all population

All naturally occurring
Anti-PP1P^k and anti-P are implicated in early spontaneous abortions

Anti-P1 frequently seen in the serum of patients with hydatid cyst disease, fascioliasis (liver fluke), and bird handlers. Antibody reactivity has been shown to be neutralized by hydatid cyst fluid and pigeon egg whites and enhanced by enzyme-treated red cells.

Auto-anti-P reported in the setting of PCH, biphasic binding IgG hemolysin, and Dx with Donath-Landsteiner.

gDNA, *Genomic deoxyribonucleic acid*; PCH, *paroxysmal cold hemoglobinuria*.

(RhAG) is essential for expression of Rh antigens. The RhAG blood group system, ISBT number (030)/symbol (RhAG)/CD number (CD241) was elevated to its own blood group system in 2008. It currently contains two low prevalence (Ol^a, RhAG4) and 2 high prevalence (Dclos, DSLK) antigens. The Rh antigenic determinants are encoded by the *RHD* (D) and *RHCE* (C, E, c, e) genes, while the RhAG antigenic determinants are encoded by the *RhAG* gene [1].

The *RHD* and *RHCE* genes are both located on chromosome 1p36.11 and each consists of 10 exons distributed over 69 kbp of gDNA in opposite orientation with their 3' end facing each other. The opposite orientation and the presence of a "hairpin" formation allow homologous DNA segments in close proximity to engage in gene recombination. Collectively they encode two protein structures compromised of 416 AA that differ by approximately 32–35 AA depending on the antigenic expression of the RhCE haplotype. Both protein structures are multipass glycoproteins with each spanning the membrane 12 times with an endocellular N-terminal and C-terminal domains. Approximately, 30,000–32,000 copies of the RHD protein are found on the surface of RBCs. In total, there are 100,000–200,000 RhD and RhCE structures in combination on RBCs. The Rh proteins (RhD and RhCE) form a core complex with the RhAG glycoprotein. Similar to the Rh protein, it spans the membrane 12 times and has an endocellular N-terminal and C-terminal domains. There are approximately 100,000–200,000 copies per RBCs. Presence of the RhAG protein is essential for Rh-antigen expression. Additionally, Rh/RhAG complex interacts with band 3, GPA, GPB, LW, CD47, ankyrin, and protein 4.2. This interaction is essential for the erythrocyte membrane integrity, as noted by the presence of stomatocytes in the Rh_{null} phenotype.

ANTIGENIC DETERMINANTS

D antigen (RH1) is the immunogenic of all antigens, present of cord cells, and protease resistant. It was identified in 1939 when Levine and Stetson who observed agglutination rate of 80% in a pregnant female. This antigen has a prevalence of 85% in Caucasians, 92% in Blacks, and 99% in Asians/Native Americans. The D antigen is unique in that it is not derived from an amino acid polymorphism but rather from the presence of the entire RhD protein. Approximately, 10–30,000 D antigenic sites are present on the surface of RBCS but due to the presence of more than 275 *RHD* alleles, numerous qualitatively and quantitatively antigenic expressions of D exist which can be organized into four groups (weak D, partial D, D_{el}, and nonfunctional RHD). Weak D (formerly known as D^u) is has a reduced amount of D antigens. Those who type weak for the D antigen have <100–10,000 D antigenic sites per RBC. This is believed to be due to SNP that induce AA (in the intracellular or transmembrane region) that may affect insertion of the RhD protein into the RBC membrane. Currently, 84 different types of weak D have been characterized out of which, type1 is the most common. Types 1–3, represent 90% of weak D types in persons of European ethnicity. Partial D types, as the name implies, are missing (or have gained) epitopes in comparison to the reference D antigen. This is the result of hybrid gene, in which portions of *RHD* are replaced by *RHCE*. As a result, these individuals are compatible of forming an anti-D when exposed to D positive RBCs with different epitopes. DIV is the most common form of partial D. D_{el} (also known as "D-elution") expresses the least amount of the D antigen, which is not detectable by routine serologic methods. Typically identified by adsorption-elution or genotyping, D_{el} is more frequently in Asian populations and accounts for 30% of the Asians population that types D negative. The nonfunctional RHD types cannot produce a full-length polypeptide and produces the D-negative phenotypes. This is most commonly observed in the European population at a rate of 15%–17%. It is due to an entirely deleted *RHD* gene. In those of African descent the nonfunctional nature of the *RHD* gene is due to 37 bp insertion into *RHD* (pseudogene) gene that results in a premature stop codon and in the Asian population, an intact but inactivation of the *RHD* gene, is typically inherited. This is extremely rare with an incidence of <0.1. Elevated D has enhanced expression of the D antigen. They typically express 75,000–200,000 D antigens per RBC. It is due to partial replacement of the RHCE gene by the RHD gene. In addition to enhanced expression of the D antigen, these individuals also express no, weak, or altered C/c and E/e antigens.

Like the *RHD* gene, numerous alleles of the *RHCE* exist. The common RHCE alleles are designated as *RHCE*01* (encodes ce), *RHCE*02* (encodes Ce), *RHCE*03* (cE), and *RHCE*04* (CE). The C antigen (RH2), which is due to inheritance of *RHCE*02 or RHCE*04*, was identified in 1941. It is due to Thymine substitution at bp 307 in exon 2 which results in a serine AA at position 103. Antithetical to the c (RH4) antigen, it has a prevalence of 68% in Caucasians and 27% in Blacks. It is expressed on cord cells and protease resistant. E antigen (RH3), which is due to inheritance of *RHCE*03 or RHCE*04*, was identified in 1943. It is due to cytosine substitution at bp 676 in exon 5 which results in a proline AA at position 226. Antithetical to the e (RH5) antigen, it has a prevalence of 29% in Caucasians and 22% in Blacks. It is expressed on cord cells and protease resistant. The c antigen (RH4), which is due to inheritance of *RHCE*01 or RHCE*03*, was identified in 1941. It is due to cytosine substitution at bp 307 in exon 2 which results in a proline AA at position 103. Additionally, a second proline at position 102 is also required for the expression of the c antigen. Antithetical to the C (RH2) antigen, it has a prevalence of 80% in Caucasians and 98% in Blacks. It is expressed on cord cells and protease resistant. The e antigen (RH5), which is due to inheritance of *RHCE*01 or RHCE*02*, was identified in 1945. It is due to guanine substitution at bp 676 in exon 5 which results in an alanine AA at position 226. Antithetical to the E (RH3) antigen, it has a prevalence of 98% in Caucasians and 98% in Blacks. It is expressed on cord cells and protease resistant.

Uncommon allelic variants of the *RHCE* gene results in altered, weak, or loss of the principle antigens. Partial C and e are common with the majority being in those of African descent.

Compound antigens also exist in the Rh blood group system. The most commonly observed ones are G and f. The G (RH12) was identified in 1958 in a D/C negative donor whose cells were agglutinated by anti-C and D. This is due to thymine substitution at bp 307 in exon 2 of *RHD* or *RHCE* which results in a serine AA at position 103 on Rh proteins expressing C and D. It has a prevalence of 84% in Caucasians and 92% in Blacks. It is expressed on cord cells and protease resistant. The f antigen (RH6) was identified in 1953. The molecular basis of this antigen is unknown but studies have shown that the c and e in cis (meaning expressed on the same haplotype, Ie R1r, or R0R0) is needed for its expression. It has a prevalence of 65% in Caucasians and 92% in Blacks. It is expressed on cord cells and protease resistant.

PHENOTYPE

The Rh antigens are inherited as a genetically linked group known as a haplotype. There are two potential options for the expression of the D antigen (D-positive or D-negative) and four potential options for the expression of the antigens expressed by the RHCE gene (ce, Ce, cE, and CE). Depending on the constellation of alleles inherited, eight potential haplotypes (Wiener's haplotypes) exist. However, only four combinations occur with significant frequency. These are known as the big four and represent ~97% of haplotypes observed. One codominant haplotype is inherited from each parent. Thus, phenotypic expression of the Rh antigens is dependent on the combined antigenic determinants in the collective haplotype (i.e., rr phenotypes are D negative and express the c/e antigens). Phenotypic expressions of the Rh haplotypes are summarized in Table 5.11.

RBCs lacking all Rh antigens are uncommon and designated as Rh_{null}. This phenotypic expression is characterized by the presence of stomatocytes and mild anemia suggesting that the Rh proteins play a role in the structural integrity of the RBC. This is more commonly due to change in the RhAG protein (known as "regulator type") which is believed to play a role in trafficking RhCE and RhD to the membrane of the RBCs. Less frequency; it is due to collective nucleotides changes in *RHD* and *RHCE* genes that prevent expression of these antigens, so called "amorph" type.

As the Rh antigens are highly immunogenic, in order of decreasing immunogenicity D > c > E > C > e, thus alloantibody formation is commonly encountered. The Rh alloantibodies are IgG, do not activate complement, require exposure, and are clinically significant as such antibodies meaning have been implicated in HDFN and hemolytic transfusion reactions. In particular, Anti-D and Anti-c are most frequently implicated in HDFN. To circumvent the formation of anti-D in Rh negative pregnant females, RhoGAM is often given at 28 weeks, at delivery, and instances of maternal fetal hemorrhage. Since alloanti-D formation can occur in the setting of partial D, RhoGAM is also indicated in this instance. If an anti-D is already formed RhoGAM administration is not warranted [1–4]. Major features of RH blood group are summarized in Table 5.12.

KIDD SYSTEM

The Kidd blood group system, ISBT number (009)/symbol (JK) was first discovered in 1951 by Allen in a patient called Mrs. Kidd who during her pregnancy produced antibodies targeted against a then unknown red cell antigen causing hemolytic anemia of the newborn. The Kidd blood group was named in memory of Mrs. Kidd's lost child, John Kidd. It currently contains 3, two polymorphic (Jk^a and JK^b) and one high prevalence (Jk3), antigens. These antigenic determinants are encoded the Kidd (*SLC14A1 or HUT11A*) gene [1].

Table 5.11 Phenotypic Expression of the Rh Haplotypes

Wiener's Haplotypes	R1: DCe	R2: DcE	R0: Dce	Rz: DCE	r': dCe	r''dcE	r: dce	r^y: dCE
Big Four	Whites: R1 > r > R2 > R0							
	Blacks: R0 > r > R1 > R2							
	Asian: R1> R2> r > R0							

Table 5.12 Major Features of RH Blood Group System: ISBT Number (004)/ Symbol (RH)

Antigens	52: D, C, E, e, F, Ce, G hr^s, C^G, RH26 (c-like),cE, hr^B, Rh41, C^w, C^x, V$^\wedge$, E^w, VS$^\wedge$, CE, D^w, hr^h, Go^a, Rh32, Rh33, Rh35, Be^a, Evans, Tar, Crawford, Riv, JAL, STEM, FPTT, BARC, JAHK, DAK$^\wedge$, LOCR, CENR, Hr$_0$, Hr, Rh29, H^h, Rh39, Nou, Sec, Dav, MAR, CEST, CELO, and CEAG
	Weak D has a reduced amount of D antigens; <100 to 10,000 D antigenic sites per RBC. Due to SNP that induce AA (in the intracellular or transmembrane region).
	Partial D types are missing (or have gained) epitopes. Due to hybrid gene of *RHD* and *RHCE*. D$_{el}$ express the least amount of the D antigen, not detectable by routine serologic methods. Nonfuctional *RHD* types cannot produce a full-length polypeptide →D-negative phenotypes
Gene (RHD OR RHCE)	Located on chromosome 1p36.11
	Both gene consists for 10 exons distributed over 69 kbp of gDNA in opposite orientation with their 3' end facing each other
	Reference *RHD* gene *encodes* D. 3 common Reference *RHCE* exist: *RHCE*01* (encodes ce), *RHCE*02* (encodes Ce), *RHCE*03* (cE), and *RHCE*04* (CE)
Protein (RhD, RhCE)	Two protein structures composed of 416 AA that differ by approximately 32–35 AA depending on the antigenic expression of the RhCE halotype.
	Both protein structures are a multipass glycoproteins. They each span the membrane 12 times with an endocellular N-terminal and C-terminal domains
	Rh/RhAG complex interacts with band 3, GPA, GPB, LW, CD47, ankyrin, and protein 4.2. This interaction is essential for the erythrocyte membrane integrity.
Antigenic Frequencies	D: 85% in Caucasians, 92% in Blacks, and 99% in Asians/Native Americans
	C: 68% in Caucasians and 27% in Blacks
	c: 80% in Caucasians and 98% in Blacks
	E: 29% in Caucasians and 22% in Blacks
	e: 98% in Caucasians and 98% in Blacks
	G: 84% in Caucasians and 92% in Blacks
	f: 65% in Caucasians and 92% in Blacks
Phenotypes	See Table 5.11
Antibodies	Rh antigens are highly immunogenic: Decreasing immunogenicity D > c > E > C > e
	IgG, do not activate complement, require exposure, and are clinically significant
	Anti-D and Anti-c are most frequently implicated in HDFN
	RhoGAM is given at 28 weeks, at delivery and instances of maternal fetal hemorrhage
	If an anti-D is already formed RhoGAM administration is not warranted
	Partial D can still form an allo-anti-D

gDNA, *Genomic deoxyribonucleic acid.*

The *SLC14A1* (solute carrier family 14, member 1) gene is located on chromosome 18q12.3. It consists for 11 exons distributed over 30 kbp of gDNA, of which only exons 4 to 11 encode the mature protein. The transcribed protein is a multipass glycoprotein which spans the membrane 10 times and has endocellular N- and C-terminal domains. First isolated in 1987, it is a 43 kDa, 389-amino acid protein with *N*-glycosylation on the third extracellular loop at ASN211. This *N*-glycosylation carrier represents 1% of the ABO antigens found on RBCs. There are approximately 14,000 copies of the Kidd protein which are found on the surface of RBCs and it has also been expressed on the kidney (vasa recta), brain, heart, pancreas, prostate, bladder, testes, and colon. The primary function of this protein, which was first discovered in a Samoan man with aplastic anemia, is urea transporter. Studies have shown, that in RBCs devoid of the Kidd antigens [Jk(a−b−)] urea crosses the membrane about 1000 times slower than in normal RBCs, and as result, these cells are more resistant to lysis in 2 M urea (2 vs. 15 min) and those with the null phenotype have a decreased ability to concentrate the urine [1–5].

ANTIGENIC DETERMINANTS

The reference allele of the JK gene (*JK*02* or *JK*B*) encodes JK2 (Jk^b) and Jk3. Numerous alleles of the JK gene exist. Alternation of the reference gene results in the presence of JK1 (Jk^a), Jk$_{null}$ [Jk: −3; Jk (a−b−)] or weak or partial expression of Kidd antigens. The two main codominant alleles, Jk^a and Jk^b, differ by a single nucleotide at position 838 (adenine and, respectively, guanine) in exon 9 and they likewise encode Jk^a and Jk^b antigens that differ by a single amino acid at residue 280 (aspartate and asparagine, respectively) on the fourth extracellular loop. The origin of the JK^3 antigen is unknown but those with the Jk (a−b−) phenotype do not typically express this antigen, the exception being those with the *In (Jk)* mutation.

Jk^a antigen was identified in 1951 in the serum of Mrs. Kidd and it has a prevalence of 77% in Caucasians, 92% in Blacks, and 72% in Asians. Jk^a is present on cord cells and is resistant to all proteolytic enzymes. The Jk^B antigen was identified in 1953 in England and has a prevalence of 74% in Caucasians, 49% in Blacks, and 76% in Asians. Similar to Jk^a, it is present on cord cells and resistant to all proteolytic enzymes. The origin of the Jk^3 epitope is unknown but its antigenicity was first observed in 1959 in a Filipino woman who became jaundiced after a blood transfusion that was possibly immunized during one of her previous pregnancies. Jk3 epitope is expressed in 100% of most populations and >99% of Polynesians (0.27%) and Finns (0.03%). It is expressed on all cells phenotypically positive for the Jk^a or Jk^b antigens, the exception being Jk (a−b−) of the *In(Jk) type*. Similar to the Jk^a and Jk^B antigens, it is expressed on cord cells and resistant to all proteolytic enzymes.

PHENOTYPE

As alluded to earlier alternations of the reference gene results numerous constellations of Kidd antigens on the surfaces of RBCs. The four main phenotypes overserved include Jk(a+b+), Jk (a−b+), Jk (a+b−), and Jk (a−b−). Major phenotypic frequencies of Kidd antigens are summarized in Table 5.13.

Table 5.13 Major Phenotypic Frequencies of Kidd Antigens

Phenotype	Caucasians	Blacks
Jk (a+b+)	50%	41%
Jk (a−b+)	23%	8%
Jk (a+b−)	26%	51%
Jk (a−b−)	Rare	Rare

The Jk (a+b+) phenotype is due to the Inheritance both the reference allele and *JK*01* which differ by one amino acid at position 280. It is the most frequency observed phenotype in Caucasians. Likewise, Jk (a−b+) and Jk (a+b−) phenotypes are due to inheritance of either one of the codominant alleles. In general, those who are homozygous for *JK*A* have been found to have higher levels of total cholesterol compared to individuals who are homozygous *JK*B*. No differences in HDL cholesterol or triglycerides levels were reported. Depending on which allele is inherited, individuals exposed to the opposite antigen may form alloantibody. Both anti-Jk^a and Jk^b shows dosage, meaning homozygous cells react more strongly with the antibody than with heterozygous cells, and have been implicated in renal transplant rejection. JK_{null} [Jk: −3; Jk (a−b−)] can arise from two different mechanisms, inheritance of an inhibitor gene (*In(Jk)*) or various SNP mutations resulting in truncation/ nonfunctional proteins. The most common genetic mutation, known as the Polynesian mutation, is due to the loss of exon 6 and does not result in the production of a Kidd protein. Individuals from this genetic background have the ability to make anti-Jk^a, -Jk^b, and -Jk^3. The anti-Jk^a and -Jk^b antibodies have been shown to express dosage and deteriorate with time. In the Japanese population, the *In(Jk)* mutation is fairly common and due to inheritance of an inhibitor gene. Despite phenotyping as Jk(a−b−), absorption/elution studies show very weak expression of Jk^3. As a result, these individuals do not form anti-Jk^3. Transient conversion to the Jk(a−b−) with subsequent anti-Jk^3 has also been documented. Weak expression of the Kidd antigens (modified expression profile) has also been demonstrated and could be the result of any multiple mutated alleles JK*A and JK*B genes. The potential antibody profile varies depending on the genetic mutation [1,2]. Major aspects of Kidd blood group system are summarized in Table 5.14.

MNS BLOOD GROUP SYSTEM

The MNS blood group system, ISBT number (002)/symbol (MNS)/CD number 235A (GPA) and 235B (GPB) was first discovered in 1927 by Landsteiner and Levine. It currently contains 46 antigens including four polymorphic (M, N, S, s), 9 high prevalence (U, En^a, ENKT, "N", ENEP, ENEH, ENAV, ENDA, and ENEV) antigens, and 33 low prevalence antigens (He, Mi^a, M^c, Vw, Mur, M^g, Vr, M^e, Mt^a, St^a, Ri^a, Cl^a, Ny^a, Hut, Hil, M^v, Far, s^D, Mit, Dantu, Hop, Nob, Or, DANTE, TSEN, MINY, MUT, SAT,

Table 5.14 Major Features of Kidd Blood Group System: ISBT Number (009)/Symbol (JK)

Antigens	3: Jk^a, Jk^b, and Jk^3
	The antigen determinants reside on the Kidd protein
	Sequence (Jk^a and Jk^b) epitopes. Origin of Jk^3 unknown, on all Jk^a and Jk^b positive cells.
Gene (SLC14A1)	Located on chromosome 18q12.3
	Consists for 11 exons distributed over 30 kbp of gDNA.
	Only exons 4–11 contribute to the mature protein
	Reference: Duffy gene (*JK*02*) encodes *Jk^b* and Jk^3
	Duffy has two antithetical antigens, JKA and JKB, which result from a SNP (838 A→G)
	JK$_{null}$ [Jk: −3; Jk (a−b−)] can arise from two different mechanisms, inheritance of an inhibitor gene (*In(Jk)*) or various SNP mutations resulting in truncation/nonfunctional proteins
Protein	14,000 copies of the Kidd protein are found on the surface of RBCs
	Kidney (vasa recta), brain, heart, pancreas, prostate, bladder, testes, and colon
	multipass protein that transports urea across the RBC membrane
Antigenic Frequencies	Jk^a: 77% Caucasians, 92% Blacks, 72% Asians
	Jk^b: 74% Caucasians, 49% Blacks, 76% Asians
	Jk^3: 100% most populations, >99% Polynesians
Phenotypes	See Table 5.13
Antibodies	IgG, many IgG plus IgM
	Bind complement if IgM is present
	Shows dosage
	Deteriorates with time → Delayed HTRs 2nd to anamnestic response

gDNA, Genomic deoxyribonucleic acid.

ERIK, Os^a, HAG, MARS, and MNTD. These antigenic determinants are encoded the Glycophorins A (*GYPA*) and Glycophorins B (*GYPB*) gene [1].

The *GYPA* and *GYPB* gene are located on chromosome 4q31.21. The GYPA gene consists of 7 exons distributed over 60 kbp of gDNA. The GYPA gene consists of five and one pseudo exons distributed over 58 kbp of gDNA. The transcribed protein Glycophorin A (GPA; MN sialoglycoprotein; SGPα) is single-pass sialoglycoprotein with a hydrophilic exocellular N-terminal domain composed of 72 amino acids, 15 O-glycosidically linked oligosaccharide chains, and 1 N-glycosidic chain. The hydrophobic endocellular C-terminal is composed of 36 amino acids and is associated with band 3. It is a 36 kDa, 131-amino acid protein which carries the M and N antigens. There are approximately 800,000 copies of the GPA protein found on the surface of RBCs. The transcribed protein Glycophorin B (GPB; Ss sialoglycoprotein; SGPδ) is also single-pass sialoglycoprotein with a hydrophilic exocellular N-terminal domain composed of 44 amino acids and 11 O-glycosidically linked oligosaccharide chains. The hydrophobic endocellular C-terminal is composed of eight amino acids. It is a 20 kDa, 72 amino acid protein which carries the S, s, and U antigens. There are approximately 200,000 copies of the GPB protein found on the surface of RBCs. Both proteins are also expressed on the renal endothelium and epithelium. They function as receptors for *Plasmodium falciparum* and serve as the major component contributing to the negative electrostatic charge of the RBC membrane [1–4,6].

ANTIGENIC DETERMINANTS

The reference allele of the *GYPA*01* (*GYPA*M*) *gene encodes* M (MNS1), MNS28, MNS29, MNS30, MNS40, MNS42, MNS44, and MNS45. The reference allele of the *GYPB*04* (or *GYPB*s*) gene encodes "N" (MN30) and s (MNS4). Numerous alleles of the abovementioned gene exist. Alternation of these genes results in the formation of the remaining antigens, aside from the U and the "N" antigen.

The two main codominant GPA antigens, M and N, differ by a three nucleotide at position 59, 71, and 72 (Cytosine/Thymine, Guanine/Adenine, Thymine/Guanine, respectively) in exon 2 and they likewise encode M and N antigens that differ by two amino acids (serine vs. leucine at position 20 and glycine vs. glutamic acid at position 24). M antigen (MNS1) was identified in 1927 and was the first identified antigen of the MNS system and has a prevalence of 78% in Caucasians and 74% in Blacks. M is present on cord cells and sensitive to ficin and trypsin. N antigen (MNS2) was also identified in 1927 and has a prevalence of 72% in Caucasians and 75% in Blacks. N is present on cord cells, sensitive to ficin and trypsin, and the lectin *Vicia graminea* reacts against the N antigens.

The two main codominant GPB antigens, S and s, differ by a single nucleotide at position 48 (Thymine vs. Cytosine, respectively) in exon 4 and they likewise encode S and s antigens that differ by one amino acid (Methionine vs. Threonine at position 48). S antigen (MNS3) was identified in 1947 and was named after the city of Sydney where anti-S was first identified. MNS3 has a prevalence of 55% in Caucasians and 31% in Blacks. The s antigen (MNS4) was identified in 1951 and is the antithetical antigen to S. It has a prevalence of 89% in Caucasians and 93% in Blacks. Both antigens are expressed on cord cells.

The "N" antigen (MNS30) is also present on GPB. The molecular origin of this antigen is the terminal N-terminal AA sequence of GPB, more specifically AA 20–24. The antigen was named "N" to represent it structurally similarity to the N antigen on GPA, so much so that it presents M+N− individuals from forming anti-N. It is present on all RBCs except those devoid of GPB or GPB-expressing He or M^v antigens.

The U antigen (MNS5) is also present on GPB. It was discovered in 1953 and named as U because of its almost universal distribution. The molecular origin of this antigen is not completely understood but expression has been postulated to require an interaction with the RhAG. It is present on all RBCs except those devoid of GPB or GPB expressing He or M^v antigens and has a prevalence of 99.9% in Caucasians and 99% in Blacks.

PHENOTYPE

As numerous constellations of MNS antigens exist on the surfaces of RBCs. The various phenotypes of MNS antigens are listed in Table 5.15. The M+N+S+s+ phenotype is the most common phenotype expressed in the Caucasians population; whereas the M+N+S−s+ phenotype is the most common phenotypic expression in Blacks. Interestingly, there are 1.5 times more copies on GPB in S+s− then S−s+ RBCs; while S+s+ RBCs have intermediate amount of GPB. Those who are negative for S and s antigen may also be negative for U. Of those that type negative for the S and s antigens, only approximately 16% are positive for the U antigen because the U negative phenotype is usually associated with an absence of GPB. Major phenotypic frequencies of MNS antigens are summarized in Table 5.15.

The Ant-M/N alloantibodies are typically naturally occurring, cold reactive, and IgM in nature. These antibodies demonstrate dosage, and are clinically insignificant. Anti-M is more frequently seen in children as opposed to adults. It is also more common in patients with bacterial infections and in M-pregnant women who give birth to M-babies. Anti-N is rare as the "N" antigen on GPB usually prevents those who are N- from forming an anti-N. In general, anti-N is almost exclusively seen in African Americans. Auto anti-N has been identified in patients on dialysis when the equipment was sterilized with formaldehyde due to modification of the N antigen. Anti-S/s/U are usually warm reactive, IgG in nature, and clinically significant. Those with anti-S in their serum usually contain antibodies to low prevalent antigens. Anti-s reactivity can be enhanced in a pH setting of 6.0. Anti-"N" does not exist [1,2]. Major features of MNS blood group system are summarized in Table 5.16.

Table 5.15 Major Phenotypic Frequencies of MNS Antigens

Phenotype	Caucasian (%)	Black (%)
M+N−S+s−	6	2
M+N−S+s+	14	7
M+N−S−s+	8	16
M+N+S+s−	4	2
M+N+S+s+	24	13
M+N+S−s+	22	33
M−N+S−s−	1	2
M−N+S+s+	6	5
M−N+S−s+	15	19
M+N−S−s−	0	0.4

Table 5.16 Major Features of MNS Blood Group System: ISBT Number (002)/Symbol (MNS)

Antigens	46: M,N, S,s, U, En^a, ENKT, "N," ENEP, ENEH, ENAV, ENDA, ENEV, He, Mi^a, M^c, Vw, Mur, M^g, Vr, M^e, Mt^a, St^a, Ri^a, Cl^a, Ny^a, Hut, Hil, M^v, Far, s^D, Mit, Dantu, Hop, Nob, Or, DANTE, TSEN, MINY, MUT, SAT, ERIK, Os^a, HAG, MARS, and MNTD.
	The antigen determinants reside on the GPA and GPB
	M and N reside on GPA. S, s, "N," and U reside on GPB
	Sequence (M, N, "N," S, and s) epitopes. Origin of U is partially sequence; complete origin unknown. M and N differ by a three nucleotide at position 59, 71, and 72 in exon 2.
	S and s differ by a single nucleotide at position 48 in exon 4
	Enzymes generally decrease all MNS antigens except U
Gene (GYPA/GYPB)	Located on chromosome 4q31.21
	GYPA consists for seven exons distributed over 60 kbp of genomic deoxyribonucleic acid
	GYPB consists for five exons and one pseudoexon distributed over 58 kbp of genomic deoxyribonucleic acid
	AA1-19 are cleaved from the mature proteins
Protein	800,000 copies of the GPA protein are found on the surface of RBCs
	200,000 copies of the GPB protein are found on the surface of RBCs
	There are 1.5 times more copies on GPB in S+s− then S−s+ RBCs.
	S+s+ RBCs have intermediate amount of GPB
	Restricted to blood cells of erythroid origin. Also found renal endothelium and epithelium
	Function as receptors for *P. falciparum*. Serve as the major component contributing to the negative electrostatic charge of the RBC membrane
Antigenic Frequencies	M: 78% Caucasians, 74% Blacks
	N: 72% Caucasians, 75% Blacks
	S: 55% Caucasians, 31% Blacks
	s: 89% Caucasians, 93% Blacks
	"N": present on all cells except those deficient in GPB or RBCs with GPB expressing He or M^v antigen
	U:99.9% Caucasians, 99% Blacks
Phenotypes	See Table 5.15

Table 5.17 Major Phenotypes of Duffy Antigens

Phenotype	Caucasians (%)	Blacks (%)
Fy (a+b+)	49	1
Fy (a−b+)	34	22
Fy (a+b−)	17	9
Fy (a−b−)	Very rare	68

The Fy (a+b+) phenotype is due to inheritance of both the reference allele and *FY*01* which differ by one amino acid at position 42. It is the most frequency observed phenotype in Caucasians. Likewise, Fy (a−b+) and Fy (a+b−) phenotypes are due to inheritance of either one of the codominant alleles. Depending on which allele is inherited, individuals exposed to the opposite antigen may develop alloantibodies. The Fy^B is a poor immunogen and has been estimated to be 20 times less immunogenic the Fy^a. FY_{null} [FY: −3; Fy (a−b−)] can arise from numerous genetic backgrounds. The most common genetic mutation known is erythroid silent. It is due to a mutation in the promoter region of the FYB allele that abolishes the expression of the Duffy glycoprotein in RBCs, but the protein is still produced in other types of cells. As such these patients do not make anti-Fyb and rarely make anti-Fy3 and anti-Fy5. A similar mutation has also been identified in the FYA allele. Additionally, the Fy (a−b−) phenotype can also be the result of point mutation that introduces a premature stop codon into the coding sequence. This mutation results in the Duffy protein being absent from all tissues and leads to the potential to make anti-Fy^a, -Fy^B, -Fy3, and -Fy5. Weak expression of the Duffy antigens (F^x phenotype) is due to a mutation in the coding sequence 265C→T (Arg897Cys). This mutation always occurs with another genetic mutation and results in decreased expression of the Fy^B, Fy^3, and Fy^5[1,2]. Major features of Duffy blood group system are summarized in Table 5.18.

KELL BLOOD GROUP SYSTEM

The Kell blood group system, ISBT number (006)/symbol (KEL) is complex and contains many antigens that are highly immunogenic. It was first discovered in 1946 and named after the first antibody producer Mrs. Kelleher. It currently contains 34 antigens including 1 polymorphic (K), 10 low prevalence (Kp^a, Js^a, Ul^a, K17, Kp^c, K23, K24, VLAN, VONG, and KYO), and 33 high prevalence (k, Kp^b, Ku, K1, K12, K13, K14, K16, K18, K19, Km, K22, TOU, RAZ, KALT, KTIM, KUCHI, KANT, KASH, KELP, KETI, and KHUL) antigens. These antigenic determinants are encoded the *KEL* gene [1].

The *KEL* gene is located on chromosome 7q34 and consists of 19 exons distributed over 21.5 kbp of gDNA and leads to expression of Kell (CD238) glycoprotein (732 amino acids). The KEL protein (CD 238) is a single-pass, 93 kDa type II glycoprotein with 15 cysteine residues that is highly folded via disulfide bonds with

Table 5.18 Major Features of Duffy Blood Group: ISBT Number (008)/ Symbol (FY)

Antigens	5: Fy^a, Fy^b, Fy^3, Fy^5, and Fy^6
	The antigen determinants reside on the DARC protein
	Sequence (Fy^a and Fy^b) or conformational (Fy^3, Fy^5, and Fy^6) epitope
Gene (DARC or FY)	Located on chromosome 1q23.2
	Consists for two exons distributed over 1.5 kbp of gDNA
	Reference: FY gene (*FY*02 or FY*B*) *encodes* FY2 (Fy^b), FY3, FY5, and FY6
	FY has two antithetical antigens, FYA and FYB, which result from a SNP (125 G→A)
	−33T →C SNP in erythroid promoter region FYB allele → Fy (a−b−)
Protein	6000–13,000 copies of the DARC protein are found on the surface of RBCs
	On endothelial cells of capillary and postcapillary venules, the epithelial cells of kidney collecting ducts, lung alveoli, and Purkinje cells of the cerebellum
	Functions as a chemokine receptor, receptor for *Plasmodium vivax and knowlesi species* (cells devoid of the Duffy antigen are resistant to invasion), suspected to impact susceptibility to HIV, and over expression confers a better prognosis in the course of prostate cancer
Antigenic Frequencies	Fy^a: 66% Caucasians, 10% Blacks, 99% Asians
	Fy^b: 83% Caucasians, 23% Blacks, 18.5% Asians
	Fy^3: 100% Caucasians, 32% Blacks, 99.9% Asians
Phenotypes	See Table 5.17
Antibodies	IgG and IgM (rare)
	Fy^B 20 times less immunogenic the Fy^a
	Usually does not bind complement
	Shows dosage
	Moderate, delayed transfusion reaction
	Mild HDFN
	Auto anti-$Fy^{a/b}$ are rare

gDNA, *Genomic deoxyribonucleic acid.*

an endocellular N-terminal and an exocellular C-terminal domain. Approximately 3500–18,000 copies of the Kell protein are found on the surface of RBCs. In addition, the Kell protein is primarily also located in the bone marrow, fetal liver, and testes. To a lesser extent, it is also found in the brain, lymphoid organs, heart, and skeletal muscle. As a member of the Neprilysin (M13) subfamily of zinc endopeptidase, it contains a pentameric sequence (HEXXH) which is central for zinc binding and proteolytic cleavage. Specifically, it cleaves big endothelin-3 (biologically inactivated), a 41 amino acid polypeptide, at Trp21-Ile22, creating bioactive endo-thelin-3 (biologically active potent vasoconstrictor). It is also believed to play a role in cell signaling and thus impact red cell maturation and differentiation. The Kell glycoprotein contains 35 antigenic sites in a mixture of high/low frequency antigen groups. It is covalently linked by disulfide bond to 444 amino acid, 37 kDa integral membrane protein, and XK that traverses the membrane 10 times. XK glycoprotein is the product of a single gene, *XK*, located on chromosome Xp21.1 which consists of three exons is distributed over 46.3 kbp. In addition to being found on RBCs, the XK glycoprotein has also been identified in brain, muscle, and heart tissues. The function of the XK protein is unclear but probably it may play a role in transporting substances into and out of cells as a mutation of the XK protein has been associated with McLeod syndrome (hemolytic anemia, myopathy, acanthocytosis, and chorea). The XK glycoprotein expresses a single antigen, Kx [1–4,8].

ANTIGENIC DETERMINANTS

The reference allele of the *KEL* gene (KEL*02) encodes k, KEL4 (Kp^b), KEL5 (Ku), KEL7, KEL11, KEL12, KEL13, KEL14, KEL18, KEL19, KEL20 (Km), KEL22, KEL26, KEL27, KEL29, KEL30, KEL32, KEL33, KEL35, and KEL36. Alternation of the reference allele results in a Kell system haplotype with one low prevalence antigen (i.e., K/Kp^a/Js^a). Additionally, compound heterozygosity or homozygosity of alterative *KEL* alleles may lead to Kell$_{null}$ [(K$_0$), no K antigens and marked increased expression of K$_x$] or Kell$_{mod}$ (weak expression of Kell antigen and increased expression of K$_x$) phenotypic expression.

K antigen (KEL1) was identified in 1946. Antithetical to KEL2, it is a low-frequency antigen in most populations but has been noted to be as high as 25% in Arabs and has a prevalence of 9% in Caucasians, 2% in Blacks, and is rare in Asians. It differs from KEL2 by a single nucleotide at position 578 (T and C, respectively) and likewise, encode K (KEL1) and k (KEL2) antigens that differ by a single amino acid at residue 193 (Methionine and Threonine), respectively. Alternatively, the KEL1 antigen is weakly expressed when Threonine is replaced by Serine or Arginine at amino acid residue 193. It is expressed on fetal RBCs at 10 weeks gestation and cord cells, protease resistant, and sensitive to DTT.

The k antigen (KEL2) was identified in 1949 and it is a high-frequency antigen in most populations. The prevalence is 99.8% in Caucasians and 100% in Blacks. It is expressed on cord cells, detected as early as 7 weeks gestation, protease resistant, and sensitive to DTT.

Kp^a antigen (KEL3) was identified in 1957. The "p" in the antigen name stands for Penny, the first name of antibody producer. It is a low-frequency antigen and antithetical to Kp^b(KEL4) and Kp^c (KEL21). This antigen has a prevalence of 2% in Caucasians and <0.01% in Blacks. It differs from Kp^b(KEL4), the reference encoded antigen, by a single nucleotide at position 841 (T and C, respectively) and likewise encode Kp^a (KEL3) and Kp^b(KEL4) antigens that differ by a single amino acid at residue 281 (Tryptophan and Arginine, respectively). Alternatively, Kp^a differs from its other antithetical antigen Kp^c by a single amino acid at residue 281 (Tryptophan and Glutamine, respectively). Interestingly, when the allelic mutation that results in Kp^a expressed is present the remainder of the Kell system antigens are suppressed (*Cis*-modifier effect) to varying degrees. It is present on cord cells, protease resistant, and sensitive to DTT.

Kp^b antigen (KEL4) was identified in 1958. It is the reference antigen and is antithetical to Kp^a (KEL3) and Kp^c (KEL21). This antigen has a prevalence of 100% in all populations. As mention above, it differs from Kp^a(KEL3) by a single nucleotide at position 841 (C and T, respectively) and likewise encode Kp^b (KEL4) and Kp^a(KEL3) antigens that differ by a single amino acid at residue 281 (Arginine and Tryptophan, respectively), whereas Kp^b(KEL4) differs from its other antithetical antigen Kp^c (KEL21) from by a single nucleotide at position 842 in exon 8. This substitution of adenine results in Kp^c(KEL21) antigen that differ by a single amino acid at residue 281 (Glutamine). It is expressed on cord cells, protease resistant, and sensitive to DTT.

Kp^c antigen (KEL3) was identified in 1945. It is a low-frequency antigen and antithetical to Kp^a(KEL3) and Kp^b (KEL4). This antigen has a prevalence of less than 0.01% in most populations. It is expressed on cord cells, protease resistant, and sensitive to DTT.

Ku antigen (KEL5) was identified in 1957. The "u" in the antigen name stands for universal, as it is present in 100% of all populations and present anytime the Kell glycoprotein is present (absent in K$_0$ phenotypes). The molecular basis of Ku is unknown. It is expressed on cord cells, protease resistant, and sensitive to DTT.

Js^a antigen (KEL6) was identified in 1958. The "Js" in the antigen name stands for John Sutter, the first name of antibody producer. It is a low-frequency antigen and antithetical to Js^b (KEL7). This antigen has a prevalence of <0.01% in Caucasians and 20% in Blacks. It differs from KEL7 by a single nucleotide at position 1790 (C and T, respectively) and likewise encode Js^a (KEL6) and Js^b (KEL7) antigens that differ by a single amino acid at residue 597 (Proline and Leucine, respectively). It is expressed on cord cells, protease resistant, and sensitive to Dithiothreitol (DTT) and EDTA glycine-acid (EGA).

The Js^b antigen (KEL7) was identified in 1963. It is a high-frequency antigen and antithetical to Js^a(KEL6) and has a prevalence of 100% in Caucasians and 99% in Blacks. It is expressed on cord cells, protease resistant, and sensitive to DTT and EGA.

Km (KEL20) antigen was identified in 1979; the "m" denotes the association with the McLeod phenotype. The molecular basis for this antigen is unknown but it is believed to be the result of an interaction between Kx and Kell proteins. This antigen is expressed in 100% of all populations. It is expressed on cord cells, protease resistant, and sensitive to DTT. EGA sensitivity has not been noted.

Kx antigen, named in 1975, is present on all RBCs except those with the McLeod phenotype. Originally believed to be a part of the Kell blood group system, this antigen was removed from the Kell system in 1990 when it was found to residue on the Xk protein. This antigen is expressed in 100% of all populations. It is expressed on cord cells, protease resistant, and enhanced by DTT. Major features of Kell blood group are summarized in Table 5.19. Major features of XK blood group are summarized in Table 5.20.

PHENOTYPE

As alluded to earlier alternations of the reference gene results numerous constellations of Kell antigens on the surfaces of RBCs. There are many phenotypic expressions which can be classified into six main categories. Homozygosity inheritance of the reference gene results in expressions of all the high frequency antigens listed above. Heterozygosity inheritance of the reference and one abnormal gene leads to inheritance of all the high-frequency Kell genes and low-frequency antigen. Homozygosity inheritance of two abnormal genes results in expression one missing high-frequency antigens or two low-frequency antigens. In the Kell null (K_0) phenotype,

Table 5.19 Major Features of Kell Blood Group: ISBT Number (006)/Symbol (KEL)

Antigens	34: K, Kp^a, Js^a, Ul^a, K17, Kp^c, K23, K24, VLAN, VONG, KYO,k, Kp^b, Ku, K1, K12, K13, K14, K16, K18, K19, Km, K22, TOU, RAZ, KALT, KTIM, KUCHI, KANT, KASH, KELP, KETI, and KHUL. The antigen determinants reside on the KELL protein
	Sequence (K, Kp^a, Js^a, Js^b, Kp^c, k, and Kp^b) or unknown (Km) epitope
	Very immunogenic (2nd only to D antigen). 10% probability of developing anti-K s/p transfusion
	Kell antigens are destroyed by thiol-reducing agents (DTT, 2-mercapatoethanol, AET, and ZZAP) and glycine acid-EDTA. Kx is not destroyed by thiol agents
Gene (DARC OR FY)	Located on chromosome 7q34
	Consists for 19 exons distributed over 21.5 kbp of gDNA
	Reference: *KEL gene (KEL*02) encodes* k, KEL4, KEL5, KEL7, KEL11, KEL12, KEL13, KEL14, KEL18, KEL19, KEL20, KEL22, KEL26, KEL27, KEL29, KEL30, KEL32, KEL33, KEL35, and KEL36
Protein (Kell, CD238)	732AA, single pass, 93 kDa type II glycoprotein with 15 cysteine residues that is highly folded via disulfide bonds with an endocellular N-terminal and an exocellular C-terminal domain. Covalently linked by disulfide bond to XK glycoprotein
	3500–18,000 copies of the Kell protein are found on the surface of RBCs
	Fetal liver and testes; lesser amounts in other tissues including various parts of the brain, lymphoid organs, heart, and skeletal muscle
	Cleaves big endothelin-3 (biologically inactivated) creating bioactive endothelin-3 (biologically active potent vasoconstrictor).

(Continued)

Table 5.19 Major Features of Kell Blood Group: ISBT Number (006)/Symbol (KEL) (*cont.*)

Antigenic Frequencies	K: 9% Caucasians, 2% Blacks
	k: 99.8% Caucasians, 100% Blacks
	Kp^a: 2% Caucasians, <0.01% Blacks
	Kp^b: 100% in all populations
	Kp^c: <0.01% in most populations
	Ku: 100% in all populations
	Js^a: <0.01%% Caucasians, 20% Blacks
	Js^b: 100% Caucasians, 99% Blacks
	Km: 100% in all populations
Phenotypes	Six main phenotypic categories
	All high frequency antigens present +Kx, Km, and Ku
	All high frequency antigens + one low frequency + Kx, Km, and Ku
	Missing one high frequency + two low frequency + Kx, Km, and Ku
	K_o: no Kell antigens, marked increase Kx, no Km. 1 in 25,000 Whites
	K_{mod}: weak Kell antigens, Ku, ↑ Kx, and No Km
	McLeod: weak Kell antigens, Ku, and No Kx/Km
Antibodies	Most commonly encountered is antibody in the Kell blood group system is anti-K. It is the most immunogenic non-ABO antibody after anti-D, IgG in nature, and more frequently found due to transfusion as opposed to pregnancy. It is implicated in acute and delayed hemolytic transfusion reactions. Anti-K has also been shown to cause HDFN. It is particularly potent since precursor RBCs carry the K antigen which leads to a suppressive (lower bilirubin and reticulocytopenia) type of anemia. Thus, titers have not been seen to correlate with severity of disease
	K_o: anti-Ku
	K_{mod}: Anti-Ku-like
	McLeod: Anti-Km and Anti-KL

gDNA, *Genomic deoxyribonucleic acid.*

all Kell antigens are absent and the Kx protein is increased, whereas in K_{mod}, the Kell antigens are markedly decreased and Kx is moderately increased. However, in McLeod, the Kell antigens are weak and the Kx antigen is missing. Individuals with the K_0, K_{mod}, McLeod make alloantibodies to Ku, Ku-like, and Km or KL (anti-Kx +anti-Km), respectively. In the setting of a missing high-frequency antigen, the corresponding antibody can be produced upon exposure. The most commonly encountered antibody in the Kell blood group system is anti-K. It is the most immunogenic non-ABO antibody after anti-D, IgG in nature, and more frequency due to transfusion as opposed to pregnancy. It is implicated in acute and delayed hemolytic transfusion reactions. Anti-K has also been shown to cause HDFN. It is particularly

Table 5.20 Major Features of XK Blood Group: ISBT Number (019)/Symbol (XK)

Antigens	1: Kx epitope source unknown
	Kell antigens are destroyed by thiol-reducing agents (DTT, 2-mercapatoethanol, AET, and ZZAP) and glycine acid-EDTA. Kx is not destroyed by thiol agents
Gene (DARC OR FY)	Located on chromosome Xp21.1
	Consists for three exons distributed over 46.3 kbp of gDNA
Protein	44 amino acid, 37 kD integral membrane protein
	Traverses the membrane 10 times
	1000 per RBC
	Fetal liver, adult SK muscle, brain, pancreas, heart low levels in adult liver kidney, spleen
	Function of the XK protein is unclear but researchers believe that it might play a role in transporting substances into and out of cells as a mutation of the XK protein has been associated with McLeod syndrome (hemolytic anemia, myopathy, acanthocytosis, and chorea)
Antigenic Frequencies	Kx: 100% all populations
Phenotype	McLeod phenotype: Hemolytic anemia, acanthocytosis, neuromuscular disorders, no Kx
	McLeod is occasional associated with X-linked chronic granulomatous disease: +NADPH oxidase disease, organisms phagocytized but not killed. Mutations in both the CYBB and XK genes
Antibodies	See key points for Kell

gDNA, *Genomic deoxyribonucleic acid.*

potent since precursor RBCs carry the K antigen which leads to a suppressive (lower bilirubin and reticulocytopenia) type of anemia. Thus, titers have not been seen to correlate with severity of disease [1,2].

KEY POINTS

- The H blood group system contains one high prevalent (H) antigen encoded by the *H* (*FUT1*) gene. The *H* gene (*FUT1* or *FUT1*01*) is located on chromosome 19q 13.33. The primary gene product of the H gene is 2-α-fucosyltransferase, an enzyme that adds the sugar α-L-fucose on to the terminal galactose of type 2 carbohydrate precursor chains attached to protein or lipids on cells.

- The homologous gene (*Se* or *FUT2* or *FUT2*01*), located 35 kbp closer to the centromere on chromosome 19q 13.33, adds the sugar α-L-fucose on to the terminal galactose of type 1 chains, carbohydrate precursor chains, attached to protein or lipids in secretions. This is functionally referred to as being a secretor.

- The H antigen is a high-frequency antigen. It has a prevalence of 99.9% in all populations on RBCs. Its expression is partially masked if a functional A or B allele is inherited, the amount of H expression is as follows in decreasing order $O > A_2 > B > A_2B > A_1 > A_1B > H+^w$.

- The three main phenotypes observed include secretor, nonsecretor, and the H-deficient phenotypes (Bombay and Para-Bombay). Nonsecretor phenotype have at least one function H gene (*FUT1*) and do not possess any function secretor genes. As a result individuals with this phenotypic expression have H antigen present on the surface of RBCs and do not have any H antigens in secretions. Those with the secretor phenotype have the H antigen present both on the surface of RBCs and in secretions. The Bombay phenotype (*hh sese*) is due to inheritance of two mutated *FUT1* and *FUT2* genes. As a result, these individuals lack ABH antigens on RBCs and in secretions. Additionally, they naturally form anti-H. Para-Bombay (H^{+w}) individuals are classically described as those who lack RBC ABH antigens on RBCs but possess them in sections. In other words, these individuals have an inactive *FUT1* and a functional *FUT2* gene.

- The ABO blood group system contains four polymorphic (A, B, AB, and A1) antigens. These antigenic determinants are encoded by the *ABO* gene, located on chromosome 9q 34.2.

- The primary gene product of the ABO gene are not the A and B antigens but glycosyltransferases that covert that H substance to blood group A and B. The addition of *N*-acetyl-galactosamine added in $\alpha1\rightarrow3$ linkage to the subterminal galactose of the H antigen results in the formation of the A antigen, whereas addition of galactose added in $\alpha1\rightarrow3$ linkage to the subterminal galactose of the H antigen results in the formation of the B antigen. Blood group O results when a mutation of the ABO gene leads to no further modification of the H antigen.

- The A antigen is the high-frequency antigen in Caucasians. It has a prevalence of 43% in Caucasians, 27% in Blacks, and 27% in Asians. The two main subgroups are A1 and A2. Approximately 80% of blood group A individuals are A1 with nearly 20% representing A2. A1 can be distinguished from A2 by the lectin *D. biflorus*, which will agglutinate A1 but not A2 cells. The B antigen has a prevalence of 9% in Caucasians, 20% in Blacks, and 27% in Asians.

- ABO antigens can be detected on RBCS as early as 5–6 weeks of gestation and reach adult levels of expression by 2–4 years. They are resistant to all proteolytic enzymes.

- The four main phenotypes observed include O, A, B, and AB. The A phenotype can be acquired by inheritance of either two functional *ABO*A1.01* alleles (AA) or one functional *ABO*A1.01* allele and one amorphic nonfunctional allele

(AO). Likewise, the B phenotype can be acquired by inheritance of either two functional *ABO* B.01* alleles (BB) or one functional *ABO* B.01* allele and one amorphic nonfunctional allele (BO). However, the O blood group phenotype is due to homozygous or compound heterozygous amorphic ABO (OO) alleles. The AB phenotype is due to inheritance of one functional *ABO*A1.01* alleles (A) and one functional *ABO*B.01* allele. Since the ABO system is characterized by the presence of naturally occurring alloantibodies, each phenotype develops alloantibodies against the missing A and B antigens.

- B(A) phenotype is characterized by having weak expression of A antigen on group B cells. The acquired B phenotype is transient phenomena in where, group A individual demonstrates weak B expression. This is due to deacetylation of the A antigen yielding a B-like sugar and is typically seen in the setting of gastrointestinal bacterial infections.

- The Lewis blood group system contains a total of six antigens including five polymorphic (Le^a, Le^b, Le^{bH}, ALe^b, and BLe^b) and one high prevalence (Le^{ab}) antigens. These antigenic determinants are encoded the LE (*FUT3*) gene, located on chromosome 19q 13.3. The primary gene product of the *LE* gene is α (1,3/4) fucosyltransferase, an enzyme that adds the sugar fucose in an $\alpha 1 \rightarrow 4$ linkage to penultimate *N*-acetyl-glucosamine on type 1 carbohydrate precursor chains attached to protein or lipids on cells. Depending on the sequential actions of the *FUT2* and *FUT3* transferases, either the Le^a or Le^b antigen will be formed. If the *FUT2* gene is functionally active first, the Le^b antigen is generated. If the *FUT3* transferase functions prior to the action of *FUT2* then Le^a is formed.

- Lewis antigens are not intrinsic to the RBC membrane but are instead passively absorbed from of the glycoproteins in the serum. Thus, expansion of plasma during pregnancy has been shown to transiently result in a Le (a−b−) phenotype.

- The Le^b antigen is the receptor for *H. pylori* and Norwalk. Increased susceptibility to *Candida* and other uropathogenic *E. coli* infections has been seen in individuals with the Le (a−b−) phenotype.

- The three main phenotypes observed include Le(a+b−), Le(a−b+), and Le(a−b−). A fourth phenotype, Le(a+b+) is rarely observed.

- The I blood group system contains one high prevalence (I). This antigenic determinant is encoded by the *I* (*GCNT2, IGnT*) gene, is located on chromosome 6p24.2 and consists for five exons, three of which are tissue specific exons (exon 1A, 1B, and 1C). The Ii collection contains one high prevalence antigen (i). The genetic basis for this antigen is unknown.

- The primary gene product of the *GCNT2* gene is 6-β-*N*-acetylgalactosaminyltransferase.

- Autoanti-I has been associated with CHAD and pneumonia due to *M. pneumoniae*. Autoanti-i is associated with infectious mononucleosis.

- The P1PK blood group system contains one polymorphic (P1), one high prevalence (P^k), and one low prevalence (NOR) antigen. These antigenic determinants are encoded by the A4GALT gene, located on chromosome 22q13.2. The primary gene product of the *A4GALT* gene is 4-α-galactosyltransferase. This enzyme transfers a galactose to the terminal sugar of a paragloboside or a lactosyl ceramide to generate the P1 antigen or P^k, respectively.

- The GLOB collection contains one high prevalence antigen (P) and the PX2 antigen, which are encoded by the *GLOB* gene, located on chromosome 3q26.1. The primary gene product of the *GLOB* gene is 3-β-N-acetylgalact osaminyltransferase. This enzyme acts in sequential action and transfers an *N*-acetylagalactosamine to a lactosyl ceramide (P^k antigen) to form GLOB (P antigen).

- P1, P^k, P, and LKE antigens all serve as receptors for P-fimbriated uropathogenic *E.coil*. P1 and P^k are receptors for *S. suis*, the causative agent of bacterial meningitis. P^k is also the receptor for Shiga toxins the causative agent for shigella dysentery and *E. coli* associated hemolytic uremic syndrome. Additionally, P^k has been shown to be physiologic receptor involved in signal modulation of α-interferon receptor and CXCR4 (HIV co-receptors) thus, it may provide for protection against HIV-1 infection; while the P antigen has been implicated as the receptor for the Parvovirus B19 virus which is the etiologic agent of erythema infectiosum (fifth disease).

- The combined expression of P1, P^k, and P antigens results in the five phenotypic expressions of P. These include P_1, P_2, P_1^k, P_2^k, and p.

- P_1 is the most common phenotypic expression with a frequency of around 80%. This phenotype expresses all three antigens (P, P1, P^k) and does not form alloantibodies.

- P_2 is the second most common of the phenotypic expression with an incidence of nearly 20%. These individuals express only the P and P^k antigens. Thus, they can generate an anti-P1 alloantibody. Anti-P1 is a naturally occurring, IgM alloantibodies are frequently seen in the serum of patients with hydatid cyst disease, fascioliasis (liver fluke), and bird handlers. Antibody reactivity has been shown to be neutralized by hydatid cyst fluid and pigeon egg whites and enhanced by enzyme treated red cells.

- Inactivating mutations in the 4-α-galactosyltransferase result in the p phenotype which is characterized by lack of P1, P, and P^k antigen. Those with the p phenotype can form an anti-PP1P^k (aka anti-Tj^a) which is a naturally occurring, combined antibody with antigenic specificity toward P, P1, and P^k and has been associated with early spontaneous.

- Autoantibody with anti-P specificity has also been reported in the setting of PCH. PCH is typically a disease of children induced by a viral illness that presents as the sudden presence of hemoglobin in the urine typically after exposure to cold temperatures. This is due to the biphasic binding of the autoanti-P IgG antibody meaning it has a tendency to bind to red cells at colder temperatures and cause hemolysis at body temperature. This characteristic is demonstrated in the Donath-Landsteiner test.

- The Rh blood group system contains 52 antigens. The main antigenic determinants are D, C, E, e, G. They are encoded by the *RHD*(D) and *RHCE* (C, E, c, and e) genes, located on chromosome 1p36.11.
- RhAG is essential for expression of Rh antigens. The RhAG blood group system contains four, two low prevalence (Ol^a and RhAG4) and two high prevalence (Dclos and DSLK) antigens. The RhAG antigenic determinants are encoded by the *RhAG* gene.
- Rh/RhAG complex interacts with band 3, GPA, GPB, LW, CD47, ankyrin, and protein 4.2. This interaction is essential for the erythrocyte membrane integrity, as noted by the presence of stomatocytes in the Rh$_{null}$ phenotype.
- D antigen (RH1) is the immunogenic of all antigens. It has a prevalence of 85% in Caucasians, 92% in Blacks, and 99% in Asians/Native Americans. The D antigen is unique in that it is not derived from an amino acid polymorphism but rather from the presence of the entire RhD protein. Numerous qualitatively and quantitatively antigenic expressions of D exist which can be organized into four groups (weak D, partial D, D$_{el}$, and nonfunctional RHD). Weak D has a reduced amount of D antigens. Partial D types are missing (or have gained) epitopes in comparison to the reference D antigen. This is the result of hybrid gene, in which portions of *RHD* are replaced by *RHCE*. As a result, these individuals are compatible of forming an anti-D when exposed to D positive RBCs with different epitopes. DIV is the most common form of partial D. D$_{el}$ expresses the least amount of the D antigen, which is not detectable by routine serologic methods. The nonfuctional *RHD* types cannot produce a full-length polypeptide and gives rise to the D-negative phenotypes. This is most commonly observed in the European population at a rate of 15%–17%. It is due to an entirely deleted *RHD* gene. In those of African descent, the nonfunctional nature of the *RHD* gene is due to 37bp insertion into *RHD* (pseudogene) gene that results in a premature stop codon.
- The Rh antigens are inherited as a genetically linked group known as a haplotype. There are two potential options for the expression of the D antigen (D-positive or D-negative) and there are four potential options for the expression of the antigens expressed by the RHCE gene (ce, Ce, cE, and CE). Depending on the constellation of alleles inherited, eight potential haplotypes (Wiener's haplotypes) exist. However, only four combinations occur with significant frequency. These are known as the big four and represent ~97% of haplotypes observed (R1 > r > R2 > R0).
- RBCs lacking all Rh antigens are uncommon and designated as Rh$_{null}$. This phenotypic expression is characterized by the presence of stomatocytes and mild anemia suggesting that the Rh proteins play a role in the structural integrity of the RBC.
- The Kidd blood group system contains three antigens: two polymorphic (Jk^a and JK^b) and one high prevalence (Jk3). These antigenic determinants are encoded the Kidd (SLC14A1 or HUT11A) gene [1], located on chromosome 18q12.3.

- The primary function of this protein, which was first serendipitously discovered in a Samoan man with aplastic anemia, is urea transportation. Studies have shown, that in RBCs devoid of the Kidd antigens [Jk(a−b−)] urea crosses the membrane about 1000 times slower than in normal RBCs, hence these cells which are more resistant to lysis in 2 M urea (2 vs. 15 min) and those with the null phenotype have a decreased ability to concentrate the urine [1–5].
- The four main phenotypes overserved include Jk(a+b+), Jk (a−b+), Jk (a+b−), and Jk (a−b). The Jk (a+b+) phenotype is due to both the Inheritance both the reference allele and JK*01 which differ by one amino acid at position 280. It is the most frequency observed phenotype in Caucasians. Likewise, Jk (a−b+) and Jk (a+b−) phenotypes are due to inheritance of either one of the codominant alleles.
- JK$_{null}$ [Jk: −3; Jk (a−b−)] can arise from two different mechanisms, inheritance of an inhibitor gene (*In(Jk)*) or various SNP mutations resulting in truncation/ nonfunctional proteins.
- The MNS blood group system contains 46 antigens. The main antigens are M, N, S, and s. These antigenic determinants are encoded the *GYPA* and *GYPB* gene, located on chromosome 4q31.21. The GYPA gene GPA carries M and N antigens. The GYPB gene carries S, s, and U. They function as receptors for *P. falciparum* and serve as the major component contributing to the negative electrostatic charge of the RBC membrane.
- Numerous constellations of MNS antigens exist on the surfaces of RBCs. The M+N+S+s+ phenotype is the most common phenotype expressed in the Caucasians population, whereas the M+N+S−s+ phenotype is the most common phenotypic expression in Blacks.
- The Anti-M/N alloantibodies are typically naturally occurring, cold reactive, IgM in nature, demonstrate dosage, and are clinically insignificant. Auto anti-N has been identified in patients on dialysis when the equipment was sterilized with formaldehyde due to modification of the N antigen. Anti-S/s/U are usually warm reactive, IgG in nature, and clinically significant.
- The Duffy blood group system contains five antigens: two polymorphic [FY1 (Fy^a) and FY2 (Fy^b)] and three high prevalence (FY3, FY5, and FY6). These antigenic determinants are encoded the Duffy (DARC or *FY*) gene located on chromosome 1q23.2.
- The DARC gene encodes DARC protein which carries the Duffy antigens.
- DARC protein is a receptor for *P. vivax* and *knowlesi species* (cells devoid of the Duffy antigen are resistant to invasion), suspected to impact susceptibility to HIV, and over expression confers a better prognosis in the course of prostate cancer.
- The four main phenotypes overserved include Fy (a+b+), Fy (a−b+), Fy (a+b−), and Fy (a−b−). The Fy (a+b+) phenotype is due to the Inheritance of both the reference allele and *FY*01* which differ by one amino acid at position 42. It is the most frequently observed phenotype in Caucasians. Likewise, Fy (a−b+) and Fy (a+b−) phenotypes are due to inheritance of either one of the

codominant alleles. Depending on which allele is inherited, individuals exposed to the opposite antigen can result in alloantibody formation.

- FY_{null} [FY: −3; Fy (a−b−)] can arise from numerous genetic backgrounds. The most common genetic mutation is known erythroid silent. It is due to a mutation in the promoter region of the FYB allele that abolishes the expression of the Duffy glycoprotein in RBCs, but the protein is still produced in other types of cells. As such these patients do not make anti-Fyb and rarely make anti-Fy3 and anti-Fy5.
- The Kell blood group system contains 34 antigens that are highly immunogenic. These antigenic determinants are encoded the *KEL* gene, located on chromosome 7q34 and leads to expression of Kell (CD238) glycoprotein (732 AA) that carries the Kell antigens. It is a type II glycoprotein with 15 cysteine residues that is highly folded via disulfide bonds that is bound to XK protein. It cleaves big endothelin-3 (biologically inactivated) creating bioactive endothelin-3 (biologically active potent vasoconstrictor).
- XK glycoprotein is the product of a single gene, XK, located on chromosome Xp21.1. The function of the XK protein is unclear but it might play a role in transporting substances into and out of cells as a mutation of the XK protein has been associated with McLeod syndrome (hemolytic anemia, myopathy, acanthocytosis, and chorea).
- There are many phenotypic expressions which can be classified into six main categories. Homozygosity inheritance of the reference gene results in expressions of all the high frequency antigens. Heterozygosity inheritance of the reference and one abnormal gene leads to inheritance of all the high-frequency Kell genes and low-frequency antigen. Homozygosity inheritance of two abnormal genes results in expression of one missing high-frequency antigens or two low-frequency antigens. Kell null (K_0) phenotype all Kell antigens are absent and the Kx protein is increased. $K_{mod,}$ the Kell antigens, are markedly decreased and Kx is moderately increased. McLeod, the Kell antigens, are weak and the Kx antigen is missing. Individuals with the K_0, K_{mod}, McLeod make alloantibodies to Ku, Ku-like, and Km or KL (anti-Kx +anti-Km), respectively.

REFERENCES

[1] Reid ME, Lomas-Francis C, Olsson ML. The blood group antigen facts book. 3rd ed. California: San Diego; 2012. p. 27–387, 489–504, 603–613.

[2] Fung MK, Hillyer CD, and Westhoff CM. "Technical Manual", Eighteenth Edition (Fung MK edited), Bethesda, Maryland, 2014. pp 291–337.

[3] Petrides M, Stack G, Cooling L, Maes LY. Practical guide to transfusion medicine. 2nd ed. Maryland: Bethesda; 2007. p. 59–93.

[4] Harmening DM. Modern blood banking & transfusion practices. 5th ed. Philadelphia, Pennsylvania: FA Davis Company; 2005. p. 108–92.

[5] Lawicki S, Covin RB, Powers AA. The Kidd (JK) Group System. Transfus Med Rev 2017;31:167–72.

[6] Heathcote DJ, Carroll TE, Flower RL. Sixty years of antibodies to MNS system hybrid glycophorins: what have we learned? Transfus Med Rev 2011;25:111–24.

[7] Wasniowska K, Lisowska E, Halverson GR, Chaudhuri A, et al. The Fya, Fy6 and Fy3 epitopes of the Duffy blood group system recognized by new monoclonal antibodies: identification of a linear Fy3 epitope. Br J Haematol 2004;124:118–22.

[8] Lee S, Russo D, Redman C. Functional and structural aspects of the Kell blood group system. Transfus Med Rev 2000;14:93–103.

Apheresis

6

INTRODUCTION

Apheresis, therapeutic plasma exchange (TPE), and plasmapheresis are terms that often are used synonymously, but incorrectly. Apheresis is a broad term which is applicable to any procedure that involves removing whole blood from a donor or patient and separating the blood into individual components so that one particular component can be removed. For a donor, apheresis may be used to collect a particular blood component such as platelets or plasma. For a patient apheresis may be used to remove a disease provoking component of blood. After removing a blood component, the remaining blood components are then returned into the bloodstream of the patient or donor. TPE specifically refers to the removal of plasma from a patient and replacement with donor plasma, whereas plasmapheresis is removal of plasma with replacement with a fluid other than plasma such as saline or albumin.

PRINCIPLES OF THERAPEUTIC PLASMA EXCHANGE

TPE is the oldest and most commonly performed therapeutic apheresis procedure. TPE can be performed manually or automatically. Manual procedure is still widely used in pediatric population. Newer automated techniques and instruments are more efficient and user friendly. TPE is a nonpharmacological treatment that removes a large portion of patient's plasma volume after separating it from the cellular components in the blood. The removed plasma volume is concomitantly replaced with appropriate fluids [1–4].

TPE can effectively remove pathological substance along with plasma in order to reduce symptoms as well as burden of disease. As a result, TPE is capable of removing or decreasing the level of circulating antibodies, antigen–antibody complexes, cytokines, abnormal plasma proteins, cholesterol, metabolic waste products, and plasma-bound toxins as well as drugs [1–4]. The extent of the pathological substance removal during TPE depends on relative volume of the plasma removed, distribution of the pathological substance between intravascular and extravascular compartments, and degree of redistribution between compartments. TPE can be used to replace a deficient factor, as in thrombotic thrombocytopenic purpura (TTP) [1]. It can be used as a standard therapy or as an adjunct therapy in combination with medications and/or surgery.

Transfusion Medicine for Pathologists. http://dx.doi.org/10.1016/B978-0-12-814313-1.00006-X

METHODS

Apheresis may be performed by using either centrifugation or filtration technology. The procedure (TPE; extracorporeal photopheresis, ECP; cytapheresis; red cell exchange, RCE) is selected based on intended separation of particular blood components. Centrifugation separates blood elements according to specific gravity (density). Blood components in order of increasing specific gravity are plasma, platelets, lymphocytes and monocytes, granulocytes, and red cells. The densest components are collect farthest from the axis of rotation, while the least dense layer is collected closest to the axis of rotation. Red cells and granulocytes have overlapping densities, resulting in poor separation. Therefore, granulocyte collected using centrifugation method contains large number of red cells. Centrifugal technology can be either discontinuous or continuous flow. Membrane filtration technology uses membranes that are permeable to high-molecular-weight proteins but not cellular elements. In the United States and Canada, centrifugal techniques are most commonly used for TPE.

VOLUME OF PLASMA TO BE EXCHANGED

An exchange of plasma volume between 1.0 and 1.5 should remove about 63%–78% of the original pathogenic substance. The amount remaining in the plasma at the end of TPE can be calculated from the simple decay formula $y/y_0 = e^{-x}$ where y = final concentration of the substance, y_0 = the initial concentration of the substance, and x = the number of times a patient's plasma volume exchanged. However, when the procedure is extended beyond 1.5 plasma volumes, the yield flattens off. It assumes that the amount remaining of a plasma constituent as long as none is added to the plasma during the TPE procedure [3,5]. The pathogenic substances targeted for removal by TPE are also present in the interstitial fluid. The rate of transfer from the interstitial space into the plasma is slow after TPE procedure has been completed. However, by the next day, the rebound is substantial and that is the reason why multiple TPE are needed to clear out the plasma as well as the interstitial space from pathogenic substances. Larger molecules that reside intravascularly such as fibrinogen (molecular weight approximately 340 kDa), IgM (molecular weight approximately 970 kDa), and cholesterol (although cholesterol has a molecular weight of 386.6, most cholesterol are bound to lipoprotein in circulation) are effectively removed (63%–78%) during TPE and are less likely to rebound. However, smaller molecules (e.g., bilirubin) are removed less than predicted (approximately 50%). However, 30% of IgG class antibody (molecular weight approximately 146 kDa) is present in the intravascular compartment.

TYPES OF REPLACEMENT FLUIDS

There are different types of replacement fluids for TPE such as crystalloids, albumin, plasma, and cryoprecipitate-reduce plasma. The selection of replacement fluid during TPE is based on the patient's underlying coagulation status and type of the disease. Isooncotic albumin and plasma are most commonly used fluids.

PATIENT EVALUATION

Patients should be evaluated by a physician familiar with apheresis before treatment begins. The indication, type of the procedure, frequency, number of treatment, type of the replacement fluid, volume of exchange, and the goal or endpoint should be documented in the consultation note. Vascular access is an important practical aspect of TPE since some patients may need central line access.

TPE procedure may be needed on an emergency basis (treatment necessary within hours), urgent basis (within a day), or routine basis. The patient's clinical condition and diagnosis should be carefully evaluated to determine the optimal timing and duration of apheresis therapy. Appropriate medical judgment should be made to determine need and urgency of apheresis. Moreover, physician requesting apheresis should consult with the physician who will be administering apheresis.

INDICATIONS

The American Society for Apheresis (ASFA) has well-researched extensive guidelines for evaluating patients who may need apheresis [1]. Category descriptions for therapeutic apheresis include:

Category I: Disorders for which apheresis is accepted as first-line therapy, either as a primary standalone treatment or in conjunction with other modes of treatment.

Category II: Disorders for which apheresis is accepted as second-line therapy, either as a standalone treatment or in conjunction with other modes of treatment.

Category III: Optimum role of apheresis therapy is not established. Decision making should be based on need of individual patients.

Category IV: Disorders in which published evidence demonstrates or suggests apheresis to be ineffective or harmful. Institutional review board (IRB) approval is desirable if apheresis treatment is undertaken in these circumstances.

Category I indications for TPE are summarized in Table 6.1. The Grading of Recommendations Assessment, Development and Evaluation (GRADE) system is used for the quality of published evidence based on multiple publications [6,7].

Grade 1A: Strong recommendation, high-quality evidence
Grade 1B: Strong recommendation, moderate-quality evidence
Grade 1C: Strong recommendation, low-quality or very low-quality evidence
Grade 2A: Weak recommendation, high-quality evidence
Grade 2B: Weak recommendation, moderate-quality evidence
Grade 2C: Weak recommendation, low-quality or very low-quality evidence

COMPLICATIONS

Therapeutic apheresis is a safe and widely used procedure although complications may still occur. The adverse effects during apheresis occur at a rate of approximately 4%–5%, but most adverse effects are mild. World Apheresis Association

Table 6.1 Category I Indications for TPE

Disease	Indication	Types of Apheresis	Grade
Acute inflammatory demyelinating polyradiculoneuropathy/ Guillain–Barre syndrome	Primary treatment	TPE	1A
Acute liver failure		TPE-high volume	1A
ANCA-associated rapidly progressive glomerulonephritis (granulomatosis with polyangiitis; and microscopic polyangiitis)	Dialysis dependence	TPE	1A
	DAH	TPE	1C
Anti-glomerular basement membrane disease (Goodpasture's syndrome)	DAH	TPE	1C
	Dialysis independence	TPE	1B
Chronic inflammatory demyelinating polyradiculoneuropathy		TPE	1B
Focal segmental glomerulosclerosis	Recurrent in transplanted kidney	TPE	1B
Hyperviscosity in monoclonal gammopathies	Symptomatic	TPE	1B
	Prophylaxis for rituximab	TPE	1C
Liver transplantation	Desensitization, ABO incompatible living donor	TPE	1C
Myasthenia gravis	Moderate-severe	TPE	1B
	Prethymectomy	TPE	1C
N-methyl-D-aspartate receptor antibody encephalitis		TPE	1C
Paraproteinemic demyelinating neuropathies/ chronic acquired demyelinating polyneuropathies	IgG/IgA	TPE	1B
	IgM	TPE	1C
Progressive multifocal leukoencephalopathy associated with natalizumab		TPE	1C
Renal transplantation, ABO compatible	Antibody-mediated rejection	TPE/immunoadsorption	1B
	Desensitization, LD	TPE/immunoadsorption	1B
Thrombotic microangiopathy, complement-mediated	Factor H autoantibodies	TPE	2C
Thrombotic microangiopathy, drug-associated	Ticlopidine	TPE	2B
Thrombotic thrombocytopenic purpura		TPE	1A
Wilson's disease, fulminant		TPE	1C

ANCA, *Anti-neutrophil cytoplasmic antibody;* DAH, *diffuse alveolar hemorrhage;* TPE, *therapeutic plasma exchange.*

(WAA)-apheresis registry data focus on adverse events in a total of 50846 procedures and reported that usually adverse events occur during the first procedure [8]. Adverse events are graded as mild, moderate (need for medication), severe or fatal.

- Mild 2.4% (due to access 54%, device 7%, hypotension 15%, and tingling 8%)
- Moderate 3% (tingling 58%, urticaria 15%, hypotension 10%, and nausea 3%)
- Severe 0.4% (syncope/hypotension 32%, urticaria 17%, chills/fever 8%, arrhythmia/asystole 4.5, and nausea/vomiting 4%)

The most common adverse effect of apheresis is symptomatic hypocalcemia due infusion of citrate as an anticoagulant. Hypotension was most common if albumin was used as replacement fluid, and allergic reaction are more common with plasma replacement.

Kinin system can be activated when plasma is exposed to foreign surfaces of plastic tubing or filtration devices, resulting in production of bradykinin. Infusion of plasma containing bradykinin can cause hypotension. Patients taking angiotensin-converting enzyme inhibitors (ACEI) are more prone to the hypotension because the drug blocks enzymatic degradation of bradykinin.

Bleeding as a consequence of coagulation factor removal is rare but has been reported. For the patients who are at risk, for example, immediate post-surgery, those with underlying coagulopathy or liver dysfunction, plasma may be used for replacement in total or as part of the replacement fluid at the end of the procedure. Apheresis can cause thrombocytopenia. Albumin-bound medications and medication with a low volume of distribution are removed by apheresis. This can result in subtherapeutic levels of medications unless dose and time of administration adjustments are made. High-molecular-weight medications such as immunoglobulins, antithymocyte globulin, and monoclonal antibodies have long intravascular half-lives and are also removed by apheresis.

SPECIAL SITUATIONS IN APHERESIS

There are some special situations during apheresis. These issues are discussed in this section.

CYTAPHERESIS

The purpose of a cytapheresis procedure is the removal of specific cellular components from a patient or collection of cellular components from blood of a donor. Cytapheresis may be leukocytapheresis where white blood cells (WBC) are separated from the blood. The cells could be discarded when used in reducing WBC count in treating acute leukemia or cells could be used for transfusion as in the case of granulocyte collection or the collection of hematopoietic progenitor cells (HPC). Cytapheresis may also be thrombocytapheresis (a therapeutic procedure where platelets are removed from blood and discarded to treat a thrombocythemic patient) or

Table 6.2 Indications for Cytapheresis

Disease	Indication	Types of Apheresis	Category	Grade
Hereditary hemochromatosis		Erythrocytapheresis	I	1B
Polycythemia vera; erythrocytosis	Polycythemia vera	Erythrocytapheresis	I	1C
	Secondary erythrocytosis	Erythrocytapheresis	III	
Hyperleukocytosis	Symptomatic	Leukocytapheresis	II	1B
	Prophylactic or secondary	Leukocytapheresis	III	2C
Thrombocytosis	Symptomatic	Thrombocytapheresis	II	2C
	Prophylactic or secondary	Thrombocytapheresis	III	2C
Psoriasis		Lymphocytapheresis	III	2C

erythrocytapheresis [red blood cells (RBCs) are removed from whole blood in a patient for therapeutic purpose or equivalent of 1 to 2 units of RBC are removed from a donor to produce RBCs for transfusion] [9]. Indications for cytapheresis are listed in Table 6.2. In acute leukemia, high blast counts (typically >100,000/µL) can result in microvasculature stasis with headache, mental status changes, visual disturbances, or dyspnea. Patients with acute or chronic lymphocytic leukemia may tolerate higher blast counts than patients with myelogenous leukemia. Massive thrombocytosis, typically >1,000,000/µL, can occur in essential thrombocythemia, polycythemia vera, or as a reactive phenomenon. Such patients may be at risk of thrombosis or hemorrhage and may be subjected to cytapheresis.

EXTRACORPOREAL PHOTOPHERESIS

ECP is also known as photopheresis and extracorporeal photochemotherapy. ECP may also be considered as a type of leukocytapheresis [9]. ECP is a cell-based immunomodulatory therapy which involves collecting the patient's mononuclear cells (MNCs) from peripheral blood. These cells are exposed to a standardized dose of photosensitizing agent, 8-methoxypsoralen (8-MOP), and are then treated with ultraviolet A (UVA) radiation, after which they are reinfused into the patient's circulation. This procedure results in cross-linking of pyrimidine bases in DNA, thus producing massive apoptosis of the treated cells. The 8-MOP was originally given orally before the apheresis. Later it is added directly to the MNC concentrate just before the UVA irradiation. This method lowers the dose of 8-MOP by 100-fold, thereby reducing side effects. The procedure was developed for use in treating cutaneous T-cell lymphoma and now extended to other diseases as in Table 6.3. ECP is typically performed on 2 consecutive days at 2–4 week intervals [1,10].

Table 6.3 Indications for Extracorporal Photopheresis

Disease	Indication	Category	Grade
Atopic (neuro-) dermatitis (atopic eczema), recalcitrant		III	2C
Cardiac transplantation	Cellular/recurrent rejection	II	1B
	Rejection prophylaxis	II	2A
Cutaneous T-cell lymphoma; mycosis fungoides; Sézary syndrome	Erythrodermic	I	1B
	Nonerythrodermic	III	2C
Graft-versus-host disease	Skin (chronic)	II	IB
	Non-skin (chronic)	II	IB
	Skin (acute)	II	1C
	Non-skin(acute)	II	1C
Inflammatory bowel disease	Crohn's disease	III	2C
Lung transplantation	Bronchiolitis obliterans syndrome	II	1C
Nephrogenic systemic fibrosis		III	2C
Pemphigus vulgaris	Severe	III	2C
Psoriasis		III	2B
Scleroderma (systemic sclerosis)		III	2A

HEMATOPOIETIC PROGENITOR CELL COLLECTION

HPC products are used during hematopoietic stem cell transplantation. Hematopoietic stem cells are mobilized from bone marrow into peripheral blood by treatment with hematopoietic growth factors. Granulocyte-colony stimulating factor (G-CSF) is used primarily. Plerixafor (AMD3100) is an antagonist of CXCR4 and used in combination with G-CSF in the patients who have poor mobilization. A CD 34+ cell dose of $\geq 2 \times 10^6$/kg recipient weight generally results in satisfactory engraftment. HPC apheresis collections are long and extensive apheresis procedures, with 15–30 L of a patient's blood processed over 4–6 h. The prolong infusion of citrate anticoagulant is associated with adverse effect of hypocalcemia.

LOW-DENSITY LIPOPROTEIN APHERESIS

Familial hypercholesterolemia (FH) is an autosomal dominant disorder due to mutations of hepatocyte apolipoprotein-B (apo-B) receptors producing decreased hepatic low-density lipoprotein (LDL) removal. Homozygous of FH usually show highly elevated cholesterol levels of 650–1000 mg/dL (desirable cholesterol level <200 mg/dL) and xanthomata by age 4 year, and death from coronary heart disease by age 20. Heterozygotes exhibit cholesterol of 250–550 mg/dL, xanthomata by age 20, and atherosclerosis by age 30. More recently, gain-of-function mutations in proprotein

Table 6.4 Special Indication

Disease	Indication	Types of Apheresis	Category	Grade
Age-related macular degeneration, dry		Rheopheresis	I	1B
familial hypercholesterolemia	Homozygotes	LDL apheresis	I	1A
	Heterozygotes		II	1A
Focal segmental glomerulosclerosis	Steroid resistant in native kidney	LDL apheresis	III	2C
Lipoprotein (a) hyperlipoproteinemia		LDL apheresis	II	1B
Peripheral vascular diseases		LDL apheresis	II	1B
Phytanic acid storage disease (Refsum's disease)		LDL apheresis	II	2C

LDL, *Low-density lipoprotein.*

convertase subtilisin/kexin type 9 (PCSK9) have been identified that result in famil-ial autosomal dominant hypercholesterolemia, a disease characterized by elevated LDL-C concentration. Recently two PCSK9 inhibitors; evolocumab, and alirocumab are available. These inhibitors are monoclonal antibodies that dramatically reduce LDL cholesterol up to 60% [11].

LDL apheresis is the selective removal of LDLs from the blood with the return of the remaining components. A variety of instruments are available which remove LDL cholesterol based on charge (dextran sulfate and polyacrylate), size (double-membrane filtration), precipitation at low pH (heparin-induced extracorporeal LDL-precipitation, HRLP), or immunoadsorption with anti-Apo B-100 antibodies. Goal is to reduce the time-averaged LDL cholesterol by 60%, usually once every 1–2 weeks. It has been recommended that apheresis should begin by age six or seven to prevent aortic stenosis occurring in homozygous with FH [1]. Special indications for apher-esis are listed in Table 6.4.

PEDIATRIC THERAPEUTIC APHERESIS

Central venous access is required in pediatric population for apheresis because peripheral access could not accommodate the catheters or may collapse under pres-sure of the blood flow. Indications for the RBC prime include: (1) the extracorporeal volume is >15% of total blood volume in children, (2) children who weigh <20 kg regardless of the Hb level, and (3) when any degree of reduction in the circulating red cell volume is undesirable regardless of the patient's body weight and Hb level, such as patients with severe anemia, hemodynamic instability, tissue/organ ischemia, or underlying cardiopulmonary disease [5,13]. RCE is one of the most common indica-tions for pediatric apheresis.

Exchange transfusion is a manual procedure in which a patient's blood is replaced with donor's whole blood that has been resuspended to a specific hematocrit using compatible fresh frozen plasma. Indications for exchange transfusion are to prevent kernicterus by correcting hyperbilirubinemia, correcting anemia in infants with erythroblastosis, hydrops or severe sepsis, as well as preventing acute events in sickle cell disease.

RED CELL EXCHANGE

RCE is the selective removal of a patient's red cells and the simultaneously replacement with donor's red cells. Common indications of RCE are listed in Table 6.5. Special requirements for RBCs used in RCE include preferably fresh RBC units <7 days old especially for patients with acute chest syndrome, negative for sickle cell trait, leukocyte-reduced, partial phenotype-match for C, E, and K. In smaller children (body weight <20 kg) or patients with severe anemia (hematocrit <18%), there is concern about the removal of >15% of the patient's red cell mass in the machine during RCE. Such a reduction can cause sudden hypoxemia in an anemic patient. To prevent the problem, the apheresis machine can be primed with RBCs instead of normal saline, so the donor's red cells enter the return line when the patient's blood is drawn into the machine [5,12].

Table 6.5 Indications for Red Blood Cell Exchange

Disease	Indication	Category	Grade
Sickle cell disease, acute	Acute stroke	I	1C
	Acute chest syndrome, severe	II	1C
	Priapism		
	Multiorgan failure	III	2C
	Splenic/hepatic sequestration;	III	2C
	Intrahepatic cholestasis	III	2C
Sickle cell disease, non-acute	Stroke prophylaxis/iron overload prevention	I	1A
	Recurrent vaso-occlusive pain crisis	III	2C
	Preoperative management	III	2A
	Pregnancy	III	2C
Babesiosis	Severe	II	2C
Erythropoietic porphyria, liver disease		III	2C
Malaria	Severe	III	2B
Prevention of RhD alloimmunization after RBC exposure	Exposure to RhD(+) RBCs	III	2C

RBC, *Red blood cell.*

RHEOPHERESIS

Rheopheresis (also called double-filtration plasmapheresis, cascade filtration plasmapheresis, or double-membrane plasmapheresis) is a special method where plasma is first separated by centrifugation and then passed through a special filter (rheofilter) where high-molecular-weight substances (e.g., α2-macroglobulin, fibrinogen, fibronectin, LDL-cholesterol, lipoprotein (a), von Willebrand factor) are removed. Rheopheresis also results in a reduction in blood and plasma viscosity, platelet and red cell aggregation, and enhanced red cell membrane flexibility which may also improve retinal pigment epithelium perfusion and function. Currently, the filtration devices necessary for this treatment are not licensed in the United States but are available in Europe and Canada [1,13].

THERAPEUTIC PHLEBOTOMY

Therapeutic phlebotomy is the removal of whole blood to treat diseases which are associated with elevated iron stores such as hereditary hemochromatosis or RBC mass (e.g., polycythemia vera). One unit (500 mL) of whole blood contains 200–250 mg iron. Adverse effects of therapeutic phlebotomy are vasovagal reactions, bruising, or hematoma at the site of venipuncture.

THROMBOTIC THROMBOCYTOPENIC PURPURA

In TTP, deficiency of the von-Willebrand (vWF)-cleaving metalloprotease (ADAMTS-13; also known as a disintegrin and metalloproteinase with a thrombospondin type 1 motif, member 13), a zinc containing enzyme responsible for cleavage of vWF results in accumulation of high-molecular-weight vWF multimer with subsequent platelet activation and platelet-rich thrombi in the microvasculature. An inhibitor of ADAMTS-13 can be demonstrated in acquired cases. Plasmapheresis with fresh frozen plasma is first-line treatment for TTP with the goal of removing both the inhibitor and large vWF multimer while simultaneously replacing the ADAMTS-13. Cryo-reduced plasma may be used as alternative replacement fluid in patients not responsive to plasma replacement. The rational is to provide a product deficiency in the substance (e.g., von Willebrand factor) that is contributing to the sequelae of the disease. However, studies have shown equivocal results in the used of cryoprecipitate-reduced plasma and although it is an alternative replacement fluid for TTP, it is generally not used. One recent study showed that the use of cryoprecipitate-poor plasma as replacement may be associated with more frequent acute exacerbations.

KEY POINTS

- TPE is the oldest and most commonly performed therapeutic apheresis procedure. TPE can be performed manually or automatically. Manual procedure is still widely used in pediatric population.
- For TPE to be effective therapy, the pathogenic substances should be removed with the plasma in order to reduce symptoms as well as burden of disease.

TPE is capable of removing or decreasing the levels of circulating antibodies, antigen–antibody complexes, cytokines, abnormal plasma proteins, cholesterol, metabolic waste products, and plasma-bound toxins and drugs.

- TPE may be performed by using either centrifugation or filtration technology but in the United States and Canada centrifugation technology is commonly used.
- Centrifugation separates blood components according to specific gravity (density). Blood components in order of increasing specific gravity are plasma, platelets, lymphocytes and monocytes, granulocytes, and red cells. The densest component is collected farthest from the axis of rotation, while the least dense layer is collected closest to the axis of rotation.
- Membrane filtration technology uses membranes that are permeable to high-molecular-weight proteins but not cellular elements.
- An exchange of plasma volume between 1.0 and 1.5 would remove about 63%–78% of the original pathology substance.
- There are different types of replacement fluids for TPE such as crystalloids, albumin, plasma, and cryoprecipitate-reduce plasma.
- Category I: disorders for which apheresis is accepted as first-line therapy, either as a primary standalone treatment or in conjunction with other modes of treatment.
- The extent of adverse side effects during apheresis occurs at a rate of approximately 4%–5 %, but the majorities of such adverse effects are mild.
- Mild adverse effects: 2.4% (due to access 54%, device 7%, hypotension 15%, and tingling 8%)
- Moderate adverse effects: 3% (tingling 58%, urticaria 15%, hypotension 10%, and nausea 3%)
- Severe adverse effects: 0.4% (syncope/hypotension 32%, urticaria 17%, chills/fever 8%, arrhythmia/asystole 4.5, and nausea/vomiting 4%)
- Patients taking ACEI are more prone to the hypotension because the drug blocks enzymatic degradation of bradykinin.
- Hematopoietic stem cells are mobilized from bone marrow into peripheral blood by treatment with hematopoietic growth factors. G-CSF is used primarily. Plerixafor (AMD3100) is an antagonist of CXCR4 and used in combination with G-CSF in the patients who have poor mobilization.
- Special requirements for RBCs used in RCE: preferably fresh RBC units <7 days old especially for patients with acute chest syndrome, negative for sickle cell trait, leukocyte-reduced, partial phenotype-match for C, E, and K.
- In TTP, a deficiency of the vWF-cleaving metalloprotease ADAMTS-13 results in accumulation of high-molecular-weight vWF multimers with subsequent platelet activation and platelet-rich thrombi in the microvasculature. An inhibitor of ADAMTS-13 can be demonstrated in many cases. Plasmapheresis with fresh frozen plasma is first-line treatment for TTP with the goal of removing both the inhibitor and large vWF multimer while simultaneously replacing the ADAMTS-13.

REFERENCES

[1] Schwartz J, Padmanabhan A, Aqui N, Balogun RA, et al. Guidelines on the use of therapeutic apheresis in clinical practice–evidence-based approach from the writing committee of the American Society for Apheresis: the seventh special issue. J Clin Apher 2016;31:149–338.

[2] Bramlage CP, Schroder K, Bramlage P, Ahrens K, et al. Predictors of complications in therapeutic plasma exchange. J Clin Apher 2009;24:225–31.

[3] Ward DM. Conventional apheresis therapies: a review. J Clin Apher 2011;26:230–8.

[4] Sanchez AP, Ward DM. Therapeutic apheresis for renal disorders. Semin Dial 2012;25:119–31.

[5] Jeffrey W. Therapeutic apheresis a physician's handbook. 3rd ed. Maryland: American Association for Blood Banks (AABB) Bethesda; 2011.

[6] Guyatt GH, Oxman AD, Vist GE, Kunz R, et al. GRADE: an emerging consensus on rating quality of evidence and strength of recommendations. BMJ 2008;336:924–6.

[7] Jaeschke R, Guyatt GH, Dellinger P, Schunemann H, et al. Use of GRADE grid to reach decisions on clinical practice guidelines when consensus is elusive. BMJ 2008;337:a744.

[8] Mörtzell Henriksson M, Newman E, Witt V, Derfler K, et al. Adverse events in apheresis: an update of the WAA registry data. Transfus Apher Sci 2016;54:2–15.

[9] Winters JL. Plasma exchange: concepts, mechanisms and an overview of American Society for Apheresis guidelines. Hematol Am Soc Hematol Educ Program 2012;2012:7–12.

[10] Klassen J. The role of photopheresis in the treatment of graft-versus-host disease. Curr Oncol 2010;17:55–8.

[11] Galema-Boers AMH, Lenzen MJ, Sijbrands EJ, Roeters van Lennep JE. Proprotein convertase subtilisin/kexin 9 inhibitors in patients with familial hypercholesterolemia: initial clinical experience. J Clin Lipidol 2017;11:674–81.

[12] Sarode R, Ballas SK, Garcia A, Kim HC, et al. Red blood cell exchange: 2015 American Society for Apheresis consensus conference on the management of patients with sickle cell disease. J Clin Apher 2017;32(5):342–67. doi: 10.1002/jca.21511.

[13] Blaha M, Rencova E, Langrova H, Studnicka J, et al. Rheohaemapheresis in the treatment of nonvascular age-related macular degeneration. Atheroscler Suppl 2013;14:179–84.

Blood components: Transfusion practices

7

INTRODUCTION

This chapter focuses on indications of blood component therapy based on a review of the literature and current guidelines for transfusion of blood components. Common contraindications are discussed when clinically relevant and each section of blood component therapy also includes appropriate dosage of each component and expected appropriate response to transfusion. In addition, administration consideration and clinical judgment considerations are also discussed in each section with reflection on certain clinical consideration for specific components such as massive transfusion and platelet refractoriness.

RED BLOOD CELL TRANSFUSION

Red blood cells (RBCs) function by carrying oxygen bound to hemoglobin (Hgb) and delivering that oxygen to the body's tissues and organs. Therefore, RBC transfusions should be indicated for patient with symptomatic deficiency in oxygen carrying capacity and/or inadequate circulating RBC mass, which is considered a standard definition for RBC transfusion [1]. The indication of RBC transfusion is straightforward in patients with acute blood loss experiencing hemorrhagic shock. In addition, RBC transfusion may also be useful in patients with signs and symptoms consistent with symptomatic anemia: tachycardia, palpitations, cooling in extremities, pallor, hypotension, acidosis, increased respiratory rate, decreased urinary output, and altered mental status [2]. Additionally, RBC transfusion is indicated in certain clinical conditions, such as a red cell exchange for patients with sickle cell disease undergoing maintenance procedures to prevent iron overload/stroke prophylaxis or exchange transfusion in neonates with hemolytic disease of the fetus/newborn. However, the indication for RBC transfusion is not as simple in patients who are not experiencing the abovementioned conditions. However, more recent evidence-based guidelines have become available to help balance the need for RBC transfusion versus the risk associated with transfusion, such as transfusion-transmitted diseases and transfusion reactions.

It should be stated that RBC transfusion should be based on the clinical presentation of the patient rather than a defined Hgb and/or hematocrit (Hct) parameter. However, as a general guideline, RBC transfusion will typically be indicated for patients with Hgb less than 7 g/dL and typically not indicated for patients that have Hgb greater than 10 g/dL.

For many years, it was a standard practice to transfuse RBCs to a patient that had Hgb less than 10 g/dL. However, physiologically this is not required in the majority of patients as the body has compensatory mechanisms to maintain appropriate tissue oxygenation through either increased cardiac output and/or increased oxygen extraction [3]. Therefore, healthy individuals are likely to have a large physiologic reserve, whereas patients with comorbidities may have a limited reverse. These compensatory mechanisms must be considered for each individual patient when determining what spectrum of Hgb concentration an individual can tolerate and at what point transfusion may be beneficial.

More recently, there have been several randomized control trial investigating various populations and determining what threshold should be appropriate for RBC transfusion. One of the first major randomized control trials that addressed this question was the Transfuse Requirements In Critical Care (TRICC) trial. The TRICC trial was a large randomized trial published in 1999 where authors assigned over 800 intensive care unit patients to either a restrictive transfusion strategy (Hgb transfusion trigger of less than 7 g/dL) or a liberal transfuse strategy (Hgb transfusion trigger of less than 10 g/dL) with the primary end point of the study being 30-day mortality [4]. The results indicated that there is no significant statistical difference in 30-day mortality between the restrictive and liberal transfusion strategy. In fact, on subgroup analysis it was shown that patients with less severe illness, as defined by Acute Physiology and Chronic Health Evaluation II (APACHE II) score of less than 20, as well as patients younger than 55 years did significantly better with a restrictive transfusion strategy. However, patients with ischemic heart disease showed a trend to higher mortality in the restrictive group. Additionally, patients with cardiac events were less represented in the restrictive group as compared to the liberal transfusion group. Based on these findings, the authors concluded that a restrictive transfusion strategy is not inferior in this patient population except possibly in those patients with ischemic heart disease.

Since this landmark study, American Association of Blood Banks (AABB) has made the recommendation that hemodynamically stable hospitalized patients, including those who are critically ill, should follow a restrictive transfusion strategy with RBC transfusion being indicated when the Hgb is less than 7 g/dL [5].

Despite the findings of the TRICC trial, there are issues regarding the safety of using a restrictive transfusion strategy in patients with cardiovascular disease. To help further elucidate this concern, the Functional Outcome in Cardiovascular Patients Undergoing Surgical Hip Fracture Repair (FOCUS) trial evaluated postoperative hip fracture repair patients with known cardiovascular disease or cardiovascular risk factors who were randomized either to a restrictive transfusion strategy (transfusion for Hgb less than 8 g/dL or symptomatic patients defined as chest pain, orthostatic hypotension, tachycardia unresponsive to fluid resuscitation, or congestive heart failure) or a liberal transfusion strategy (transfusion for Hgb less than 10 g/dL). The primary end point of the study was death and inability to walk across a room unassisted at 60-day follow up [6]. The findings of FOCUS trial demonstrated no significant difference in the primary end point.

Additional studies such as the Transfusion Indication Threshold Reduction (TITRe2) trial, Transfusion Requirements After Cardiac Surgery (TRACS) trial, and a randomized trial performed at the Texas Heart Institute in Houston, Texas also compared clinical utility of restrictive versus liberal transfusion strategy (Hgb of 7.5 g/dL vs. 9 g/dL, Hct of 24% vs. 30%, and Hgb of 8 g/dL vs. 9 g/dL, respectively) in cardiac surgery patients. Collectively, these studies illustrate that a restrictive transfusion (Hgb 7.5–8 g/dL) strategy is also feasible in this patient population [7–9].

As highlighted by these major studies, AABB has the recommendation for a restrictive RBC transfusion strategy (transfusing for Hgb less than 8 g/dL) in orthopedic and cardiac surgery patients as well as those with preexisting cardiovascular disease. It is also noteworthy that there currently remains insufficient evidence regarding appropriate RBC transfusion strategies in patients with acute coronary syndrome, hematologic/oncology patients with severe thrombocytopenia who are at a higher risk for bleeding, and in patients with chronic transfusion dependent anemias. As a result a recommendation for a restrictive transfusion strategy cannot be applied to these patient populations [5].

DOSE AND ADMINISTRATIVE CONSIDERATIONS

In general, transfusion of one unit of RBCs is expected to increase Hgb by 1 g/dL or Hct by 3% in a non-actively bleeding adult patient with average body weight. In children, a typical RBC transfuse dose is 10–15 cc/kg which is expected to increase the Hgb by 2–3 g/dL or 6% increase in Hct.

ABO identical or compatible RBC blood components are appropriate for RBC transfusion (Table 7.1). Even though there may be infusion of incompatible donor plasma from a compatible RBC blood component (such as transfusion of Type O RBCs into a Type A patient), the residual incompatible plasma in the RBC product contains an insignificant amount of isohemagglutinins resulting in no major risk of hemolysis from the RBC product, even if multiple units are transfused. However, ABO identical should be used in whole blood transfusion as there is a significant amount of incompatible isohemagglutinins in non-ABO identical whole blood units that may potentially result in hemolysis.

Other than being compatible for ABO type (which is typically assessed by either immediate spin or more commonly electronic crossmatch), determination of compatibility for known history of alloantibodies and/or alloantibodies identified on the current serology workup is also important. In such cases, patients should receive

Table 7.1 Compatible RBC Transfusions in Component Therapy

Patient ABO Type	Compatible RBC Blood Components
O	O
A	A and O
B	B and O
AB	A, B, AB, and O

antigen-negative RBC units for previously identified alloantibodies or alloantibodies identified on the current workup. Example being D antigen negative (Rh negative) unit is provided to patients that have formed an anti-D alloantibody. In this case, a serologic crossmatch to the antihuman globulin (AHG) phase should be performed to ensure that the appropriately selected unit will be compatible with the patient.

In cases where the presence of underlying alloantibodies cannot be determined, such as a patient with a warm autoantibody, discussion with clinical team should be undertaken to determine the necessity for RBC transfusion as various methods can be employed to determine safe units for transfusion. Absorption studies can be performed to remove the warm component to allow the determination of alloantibodies. An alternative is to perform phenotyping or genotyping to determine appropriate antigen negative RBC units, thus eliminating the need to determine underlying alloantibodies. However, this should be discussed with the clinical team as knowledge of recent transfusion (last 3 months) and likelihood of recurring need for transfusion is vital in determining what method to perform as well as communication of anticipated delays for transfusion of safe blood products due to the additional time required for the various techniques and selection of appropriate RBC units. In these situations the risk versus benefit for transfusion must be decided on a case to case basis with the patient's physician and the transfusion medicine service.

In clinical situation where the necessity of transfusion outweighs the risk of incomplete blood bank testing, emergency release is an option. Under such circumstances, the attending physician is assuming responsibly that the emergent need for blood transfusion is lifesaving and outweighs the risk for the incomplete serology workup. In emergency release protocols, the attending physician can customize what blood components are appropriate for their patient with the understanding that the transfusion service will prepare and issue only what has been requested by the physician. In contrast, massive transfusion protocol (MTP) should be initiated in those clinical situations in which excessive blood loss in a patient necessitating continued blood product supply. Such protocol releases blood component on a fixed regimen, typically 1:1:1, [such as 6 RBCs, 6 fresh frozen plasma (FFP), 1 dose of platelets] and cannot be modified. For MTP, it is important that the attending physician understands that customization of blood components is not possible and that the blood bank will continue to prepare the next round until the MTP has been deactivated. Therefore, the attending physician should inform the blood bank when the MTP is no longer required.

PLATELET TRANSFUSION

For appropriate primary hemostasis, adequate number of functional platelets is essential. As a result, the indications for platelet transfusions are when a patient has a deficiency of platelets (thrombocytopenia) or the patient's platelets do not function properly (thrombocytopathy). For thrombocytopenia, prophylactic platelet transfusions are given to prevent spontaneous bleeding or those at risk for bleeding

Table 7.2 Platelet Transfusion Triggers

Any Patient	<10,000/µL
Patients with coagulation disorders, on heparin, with anatomical lesions, or undergoing CVC placement	<20,000/µL
Patients with active bleeding, undergoing lumbar puncture or major non-neuraxial surgery	<50,000/µL
Patient with bleeding into confined anatomical spaces such as intracranial, ophthalmic, and pulmonary	<100,000/µL

CVC, *Central venous catheter.*

while therapeutic platelet transfusions are used to treat active bleeding. Several studies have examined the use of platelet transfusion in these contexts and the guidelines for platelet transfusions as well as generally adopted thresholds are discussed (Table 7.2).

Prophylactic platelet transfusion is indicated when the platelet count is less than 10,000/µL. Several studies have shown that in patients with treatment-induced hypoproliferative thrombocytopenia, the use of prophylactic platelet transfusions is associated with less bleeding when transfused for a platelet count less than 10,000/µL while a higher thresholds were not associated with a significant difference in the degree of bleeding [10]. Additionally, it has also been illustrated that prophylactic platelet transfusions should be considered in patients on heparin anticoagulation, those with coagulation disorders, or anatomical lesions when the platelet count is below 20,000/µL [11]. More recently, AABB has recommended prophylactic transfusion of platelets for those undergoing placement of a central venous catheter when the platelet count is below 20,000/µL [10]. In clinical situations of active bleeding, patient that are undergoing major non-neuraxial procedures or in patients having elective lumber puncture, it is recommend to maintain the platelet count greater than 50,000/µL [10,12]. Generally, it has long been accepted to transfuse platelets to a higher threshold for anatomical locations in which bleeding into confined spaces would be detrimental. Therefore, platelet transfusions to maintain the platelet count greater than 100,000/µL are generally done for intracranial, ophthalmic, and pulmonary bleeding, although currently there is no guideline for platelet transfusion in these clinical situations, particularly for intracranial bleeding [10,2].

Platelet transfusion may also be indicated in patients with normal platelet counts but have dysfunctional platelets—thrombocytopathy. Thrombocytopathy is due to congenital conditions, such as Bernard–Soulier syndrome, Glanzmann thrombasthenia, or platelet storage pool disorders or acquired conditions. Acquired conditions of thrombocytopathy may be due to use of certain medications such as aspirin or clopidogrel (two commonly used antiplatelet medications), or related to hematologic conditions such as in myelodysplastic syndrome. Platelet dysfunction has also been well documented in cardiovascular patient during surgical procedures requiring cardiopulmonary bypass. During this procedure, platelets become activated and degranulate as they circulate through the bypass machine. Additionally, the hypothermia experienced by a patient during the bypass procedure may result in platelet dysfunction [13]. Patients with renal failure are also known to have platelet dysfunction as a result of accumulation of urea—uremic platelet dysfunction.

In patients with evidence of thrombocytopathy, the clinical presentation is vital in determining the need and extent of platelet transfusion. In addition, alternatives to platelet transfusion should be considered. Administration of other agents that increase von Willebrand factor, such as cryoprecipitate (cryo) and 1-deamino-8-D-arginine vasopressin (DDAVP), may be beneficial in these situations, particularly DDAVP in patients with uremic platelet dysfunction [14].

Classic contraindications to platelet transfusion include

- Idiopathic thrombocytopenic purpura (ITP)
- Posttransfusion purpura (PTP)
- Thrombotic thrombocytopenic purpura (TTP)
- Heparin-induced thrombocytopenia (HIT)

ITP is autoimmune condition causing rapid clearance of self as well as transfused platelets due to antibodies. PTP is a transfusion reaction resulting in the formation of alloantibodies from exposure of foreign platelet antigens in the donor blood product resulting in clearance of the transfused platelets as well as the native platelets. As a result, platelet transfusions are not significantly beneficial in these conditions and management should be immunomodulation, such as intravenous immunoglobulin. RhIg administration has also been shown to be efficacious in patients that are Rh positive with intact spleen in patients with ITP [15]. TTP is typically due to an acquired deficiency in ADAMTS13 (a disintegrin and metalloproteinase with a thrombospondin type 1 motif, member 13 which is also known as von Willebrand factor-cleaving protease) which results in platelet-rich microthrombi and a classic pentad of fever, anemia, thrombocytopenia, renal failure, and neurologic abnormalities. HIT, specially type 2, is a complication of heparin therapy resulting in the development of antibodies to the heparin/platelet factor 4 complex that ultimately leads to platelet activation and formation of platelet-rich microthrombi which results in potentially life-threatening thrombotic complications. Transfusion of platelets in these conditions has been considered a contraindication [2]. However in life-threatening bleeding, it is reasonable to transfuse platelets and several studies have found platelet transfusions not to be harmful in regards to thrombotic complications in TTP patients [16,17] as well as HIT patients [18].

DOSE AND ADMINISTRATIVE CONSIDERATIONS

The amount of platelets that is considered a dose of platelets is one apheresis platelet, which is required to have at a minimum of 3.0×10^{11} of platelets. A unit of platelets (whole blood-derived platelets/platelet concentrate) is required to contain at a minimum 5.5×10^{10} platelets. Therefore, approximately 6 units of platelets, sometimes called "a six pack," would equal a dose of platelets as it would be equivalent to an apheresis platelet in regards to platelet number. The transfusion of a dose of platelets is expected to increase the platelet count by 30,000–60,000/µL in an average size adult. Therefore, one unit of platelets would be expected to increase the platelet count by 5000–10,000/µL.

In regards to the type of platelets transfused, most blood bank stock apheresis platelets as approximately 90% of platelet transfusions are from apheresis collections [2]. The main advantage of transfusing apheresis platelet over whole blood-derived platelets is that there is less infectious risk associated with apheresis platelets as such products are derived from a single donor. Additionally, apheresis platelets are indicated in patients requiring HLA-compatible and/or HPA-matched platelets such as for patients with platelet refractoriness or neonatal alloimmune thrombocytopenia or in patients requiring IgA-deficient platelets [19].

When providing platelets, it is desirable to provide ABO-matched platelets as this would provide the most efficacious increase in the platelet count from the platelet transfusion [20]. This is due to the fact that platelets contain ABO antigens and transfusion, for example, a type A platelet into a type O recipient although would be compatible in terms of the donor plasma in the recipient, the ABO incompatible platelets would not be as efficacious. However, it still remains elusive if ABO-matched platelets provide superior clinical benefit over ABO-incompatible platelet transfusions. As a result, it is not atypical for blood bank to provide incompatible platelet transfusion, including ABO plasma incompatible platelet transfusion given that there is no clear clinical benefit to ABO match platelets as well as a constrain on blood bank inventories to maintain platelets given their short shelf life. The risk of hemolysis from ABO plasma incompatible apheresis platelet transfusions is approximately 1:3000 to 1:10,000 and seen more commonly with platelet units that have a high isohemagglutinin titer level [21,22].

ABO plasma incompatible platelet transfusions can be mitigated by removal of the excess plasma either by volume reduction, washing, or the use of platelet additive solution. Additionally, identifying units with high titers of isohemagglutinin and limiting those high titer units to ABO-match transfusions can deter a possible hemolytic transfusion reaction.

Another undesirable complication of transfusing ABO-incompatible platelets is the increased risk for the development of platelet refractoriness [23]. Platelet refractoriness is seen when there is a less than expected increase in the platelet count following platelet transfusion. In order to consider platelet refractoriness, it is important to ensure that ABO match platelets are being transfused and other nonimmunogenic causes, such as fever, sepsis, splenic sequestration, drugs, and consumptive processes such as disseminated intravascular coagulation (DIC) and TTP, are excluded as the cause for the inappropriate platelet response. If platelet refractoriness is clinically suspected, a corrected count increment (CCI) should be performed 1 h after the platelet transfusion to determine the presence of refractoriness. If the CCI is determined to be less than 5000–7500 after two consecutive transfusions then an immune cause of platelet refractoriness most commonly due to HLA sensitization to class I HLA antigens is determined.

The CCI is calculated as follows:

$$CCI = \frac{Body\ surface\ area\ (m^2) \times \left[post - pretransfusion\ platelet\ count\ \times 10^{11} \right]}{Number\ of\ platelets\ transfused\ \times 10^{11}}$$

EXAMPLE:

A gentleman with a BSA of 2 m² with a platelet count of 10,000/μL was transfused with two doses of apheresis platelets with an increase in the platelet count to 24,000/μL. What is the expected CCI if measured at 1-h post transfusion? In this example, since the number of platelets is not provided, one has to assume the standard of 3×10^{11}, therefore 6×10^{11} since two doses were given.

$$CCI = \frac{2 \times [24000 - 10000] \times 10^{11}}{6 \times 10^{11}} = 4666.6 \ (approximately\ 4667)$$

Once a patient is considered platelet refractory by the CCI, it is important to determine (1) the HLA phenotype of the patient for HLA-A and HLA-B antigens and (2) determine which HLA antibodies are present which will help guide the different platelet transfusion strategies for these patients. The three treatment options for platelet refractory patients are

1. HLA matched platelets
2. HLA antigen negative for the recipient's known HLA antibodies
3. Platelet crossmatch compatible units

HLA-matched units are graded as A to D (Table 7.3). However, it is difficult to find a perfect antigen-matched HLA platelet product and most units provided by the blood bank usually end up being some degree of B grading antigen match and may possibly be ABO incompatible. If this is the case, it may be more reasonable to select units that are HLA antigen negative for the patient's known HLA antibodies (similar to what is practiced for RBC transfusion for patient with alloimmunization) as this may provide a wider pool of platelet donor to choose from and increases likelihood of ABO matched. Crossmatch-compatible platelets are done by solid phase red cell adherence assay in which the serum of the recipient is mixed with the donor platelets and

Table 7.3 Grading for HLA Matched Platelets

Grade	Type of Platelets
A	Four antigen matched
BU	Three antigens identical, one antigen unknown
B2U	Two antigens identical, two antigens unknown
B1X	Three antigens identical, one CREG
B2UX	Two antigens identical, one unknown, one CREG
C	One mismatched antigen
D	Two or more mismatched antigens

CREG, *Cross-reactive group.*

indicator red cells are used to determine crossmatch compatibility by the pattern seen on the microwell plate. One last point to remember with HLA matched platelets is that they should be irradiated to prevent transfusion associated graft versus host disease.

Rh matching is another important consideration for platelet transfusion. Although Rh antigens are not found on platelets, it is the residual RBCs in the unit that may lead to exposure to foreign Rh antigens in the recipient. However, residual RBCs is minimal in platelet products as shown by one study that demonstrated an average RBC content of 0.036 and 0.00043 mL in whole blood-derived platelets and apheresis platelets, respectively [23]. Despite the potential for alloimmunization to the D antigen, this is typically uncommon with platelet transfusion [24]. However, selection of Rh negative platelets should be done for Rh negative women of child bearing age due to the clinical significance of hemolytic disease of the fetus/newborn in future pregnancies should anti-D alloantibodies form. All attempts should be made to provide these patients with Rh negative platelets. However, in the case in which Rh positive platelets are given, RhIg should be administered to prevent the development of anti-D alloimmunization.

PLASMA TRANSFUSION

Unlike RBC and platelet transfusion, the evidence-based guidelines regarding the use of plasma transfusion are weak or lacking appropriate randomized control trials resulting in much of the indications being based on expert opinion [25].

Clinical indications for plasma transfusion are:

1. Active bleeding in the setting of multifactor deficiency
   a. Examples being patients with dilutional coagulopathy, such as from massive transfusion, patients with disseminated intravascular coagulopathy or liver disease
2. Emergency reversal of warfarin-induced coagulopathy with active bleeding
   a. Alternatively, if available, prothrombin complex concentrates (PCC) are a more practical option and a FDA indicated use for urgent warfarin reversal (see plasma-derived section)
3. Use as replacement fluid for therapeutic plasma exchange, such as patients with TTP
4. Patients with congenital factor deficiency in which there is no available factor concentrate, such as deficiency in factor V and factor XI
5. Historically, plasma was used in patients with hereditary angioedema resulting from deficiencies of C1 esterase inhibitor, however more recently, an FDA approved concentrate is available

In addition to these indications, plasma may also be requested to improve or normalize the international normalized ratio (INR) prior to patient going for invasive procedures or surgery with the premise being that an elevated INR is associated with increased bleeding and normalization of the INR will decrease the risk of bleeding. However, review of available studies show that abnormalities in the INR are not predictive of bleeding in patients undergoing procedures [26]. Coagulation studies remain a poor predictor of clinical bleeding owing to the sensitivity of coagulation studies, in which case, slight deficiencies in coagulation factors will result in prolongation of the PT/PTT but not at factor levels associated with inadequate hemostasis [27]. Furthermore, when there are mild elevations in the INR, transfusion of plasma has been shown to have little effect in improving the INR [28], due to the exponential relationship between factor levels and the INR [2]. As a result, most institutions will consider transfuse of plasma as appropriate for an INR greater than 1.5 to 2. Additionally, if plasma is used for correction of INR before an invasive procedure or surgery, a post-transfusion INR should be obtain following the transfusion before additional plasma products are ordered given the risk/benefit of plasma in this setting.

Plasma should not be used in patients as a means to provide protein in nutritionally deficient patients and in patients requiring an increase in intravascular volume. In such cases, other volume expanders are appropriate due to the risk of transfusion-transmitted diseases and possible transfusion reactions associated with plasma [1].

DOSE AND ADMINISTRATIVE CONSIDERATIONS

The typical dose for plasma is 10–20 mL/kg and is expected to increase the coagulation factors by 20%. Additionally, because of the short in vivo half-life of some of the coagulation factors, namely, factor VII, plasma should be given as close to the time of the anticipated surgeries or procedure if being used for correction of coagulation abnormalities. Of the plasma products that can be transfused, FFP contains optimal amounts of all the coagulation factors while plasma frozen within 24 h (PF24) and more so thawed plasma show reduced levels of labial in vitro factors (factor V and factor VIII). As a result, PF24 and thawed plasma are clinically equivalent to FFP in treating multifactor deficiency, such as in as massive transfusion and surgical bleeding. However, if treating single-factor deficiency, such as patients with congenital factor V deficiency, FFP should be used.

CRYOPRECIPITATE TRANSFUSION

Cryo is the by-product of thawing FFP at 6°C resulting in a product that serves as source of factor VIII, von Willebrand factor, factor XIII, and fibrinogen (factor I). Historically, cryo was used to treat patients with hemophilia A. Now that a factor VIII concentrate is commercially available, cryo is rarely used for this purpose except in those situations where factor VIII concentrate may not be available. Additionally, cryo is no longer routinely used for treatment of von Willebrand disease or factor XIII deficiency as factor concentrates are available. Cryo can also be used as a topical fibrin sealant in combination with calcium and thrombin, but more readily available commercial products have also replaced cryo in this use as well.

The indication for cryo is for the treatment of congenital or acquired fibrinogen deficiency. However, more recently a fibrinogen concentrate has been FDA approved for the treatment of bleeding in patients with congenital fibrinogen deficiency but may not be as wide available as cryo in most institutions. In addition to the treatment of congenital or acquired fibrinogen deficiencies, cryo is used to treat bleeding in renal patients with uremia and should be considered in patients that are refractory to alternative strategies for treatment of uremic platelet dysfunction, such as DDAVP [29,30]. Cryo is not indicated as a replacement product for plasma in the treatment of multifactor deficiency coagulopathy such as in patients undergoing massive transfusion or patients with DIC since there are insignificant amount of the other coagulation factors in cryo.

DOSE AND ADMINISTRATION CONSIDERATIONS

A dose of cryo is typically considered to be 10 units. Each unit of cryo can be expected to increase the fibrinogen level by approximately 5–10 mg/dL in an average size adult. Therefore a dose of cryo (10 units) would be expected to increase the fibrinogen by about 100 mg/dL, the level of fibrinogen that is considered as the threshold to maintain adequate hemostasis. In pediatric patients, dosing 1–2 units/10 kg would be expected to increase the fibrinogen level by 100 mg/dL [1].

Although cryo is typically dispensed in units of 10, the number of units needed to achieve a desired fibrinogen level can be calculated. This requires knowing the delta change wanted in fibrinogen level as well as the plasma volume of the patient and using the typical fibrinogen content of 250 mg per unit [2].

Calculation:

Number of units = Desire change in fibrinogen × plasma volume ÷ 250 mg/unit

Plasma volume = (1 − Hct) × 0.7 dL/kg × body mass or 30 dL, if weight is unknown

EXAMPLE:

An 80 kg patient with a Hct of 28% is found to have acute bleeding after coming off bypass during cardiovascular surgery and is found to have a fibrinogen level of 77 mg/dL. The requesting anesthesiologist would like to obtain a fibrinogen level of at least 200 mg/dL. How many units would be needed?

$$\text{Number of units} = \frac{(200 - 77\,\text{mg/dL}) \times \left[(1 - 0.28) \times (0.7\,\text{dL/kg} \times 80\,\text{kg})\right]}{250\,\text{mg/units}} = 19.8 \approx 20\,\text{units}$$

One last consideration of cryo is that it has a shelf life of 6 h once thawed or 4 h if units are pulled together. Therefore, it may be imperative to discuss with the clinical team if excess cryo has been ordered to ensure that the number requested is appropriate for the clinical situation and that the units will be transfused in the allotted time to prevent wastage. An alternative strategy can also be to release one dose and reevaluate by clinical condition and laboratory studies to determine the need for additional cryo.

GRANULOCYTE TRANSFUSION

The ability to collect and administer granulocytes as a therapeutic option has been around since the 1970s. However, the use of granulocytes fell out of favor due to equivocal results from it use, variable yields from collections, as well as the cumbersome collection process as more efficacious antimicrobials were emerging. However, more recently, there has been a renewed interest in granulocytes as a therapeutic option as apheresis technology has improved as well as improved granulocyte collection from donors through the addition of granulocyte colony stimulating factor, thus allowing increased mobilization of granulocytes from the donor [31].

Granulocytes should be considered as a therapeutic option when

1. The patient's absolute neutrophil count (ANC) is less than 500/μL
2. There is a bacterial or fungal infection or suspected infection, for 24–48 h
3. The patient shows lack of improvement to antimicrobial treatment
4. There is a reasonable chance that the patient will have marrow recovery

Other indications for which granulocytes may show improvement include neonates with sepsis or patients with neutrophil dysfunction, such as patients with chronic granulomatous disease.

DOSE AND ADMINISTRATIVE CONSIDERATIONS

The analysis of published studies has shown that the minimum number granulocytes that may show benefit in neutropenic patients is 1×10^{10} with improved results in patients that receive higher yields, such as pediatric patients that have a smaller blood volume [31]. As a result, standards require a dose of granulocytes to contain at least 1×10^{10} neutrophils.

Granulocytes should also be ABO compatible with the recipient as most units of granulocytes contain more than 2 mL of RBCs. Therefore, it must be treated in a similar fashion as transfusion of RBCs. The cytomegalovirus (CMV) status of the patient should also be obtained. Granulocytes cannot be leukocyte reduced because such procedure will also remove the granulocytes. Thus patients that are CMV negative should be transfused with CMV seronegative units as leukoreduced CMV safe products cannot be prepared. Granulocytes should also be irradiated to prevent transfusion-associated graft versus host disease (TA-GVHD). Granulocytes have a shelf life of 24 h; therefore they should be transfused as soon as they received in the blood bank after collection.

PLASMA-DERIVED PRODUCTS

There are various plasma-derived products which are used in transfusion medicine. These products are discussed in this section.

ALBUMIN AND PLASMA PROTEIN FRACTIONS

Albumin is derived from human plasma and contains 96% albumin and 4% globulins along with other proteins. It is prepared by cold alcohol fractionation and then it is heat inactivated to reduce the risk of viral transmission. As a result, albumin is considered extremely safe. Plasma protein fractions (PPF) are a less purified product and contain 83% albumin and 17% globulins along with other proteins. These products lack coagulation factors and come in concentrates of 25% and 5% for albumin and 5% for PPF. 5% albumin is equivalent to human plasma while 25% albumin has increase osmotic and oncotic effect. Albumin is indicated for the treatment of hypovolemia, acutely hypoproteinemic patients, burns, and as a replacement fluid for therapeutic apheresis. It should not be used in patients for correction of nutritional deficiency.

CONCENTRATES: VIII, IX, X, XIII, FIBRINOGEN (FACTOR I), PROTEIN C, AND ANTITHROMBIN

Factor VIII can be prepared from plasma as well as by recombinant methods. The plasma-derived form is manufactured through fractionation of pooled human plasma and range in the degree of purity based on the method of purification used. Those concentrates that have a lower purity may contain von Willebrand factor and can also be used for the management of patients with von Willebrand disease while those concentrates with a high purity of factor VIII are used for the management of patients with Hemophilia A. These products are virus inactivated and have a half-life in the range 8–12 h.

Factor IX can be prepared from plasma as well as recombinant methods. The plasma-derived product undergoes a virus inactivation process to reduce the risk of infectious diseases. Unlike PCC, this is a highly pure product of factor IX and contains nonclinically significant amounts of factor II, factor VII, and factor X. Factor IX concentrate is used for the management of hemophilia B and has a half-life in the range of 18–24 h.

Factor X is a plasma-derived factor X concentrated and is indicated in the management of patients with hereditary factor X deficiency. This product has a half-life of about 30 h.

Factor XIII is another plasma-derived product indicated for the treatment of factor XIII deficiency. This product has a half-life of about 9 days.

Fibrinogen concentrate is a pasteurized human-derived product and is indicated for the treatment of acute bleeding in patients with congenital fibrinogen deficiency, afibrinogenemia, and hypofibrinogenemia. The half-life of this product is approximately 3 days.

Protein C concentrate is used for the treatment of patient with congenital protein C deficiency. Currently only a plasma-derived form is available as the recombinant form was removed from the market in 2011.

Antithrombin is produced from both recombinant and plasma-derived methods. It is used in the management of patients with hereditary antithrombin deficiency. The half-life of the plasma-derived form is longer than the recombinant form (approximately 60 vs. 10 h).

PCCS, APCC, AND RECOMBINANT FACTOR VIIA

PCC is lower purity product of factor IX that contains other vitamin K dependent factor. There are two available versions of PCC approved for use in the United States that differ in the level of factor VII. The three-factor concentrate PCC contains factor II, factor IX, and factor X while the four-factor concentrate contains factor II, factor VII, factor IX, and factor X. For both products, the purity is based on the level of factor IX and they are virus inactivated to reduce the risk of infectious disease. The four-factor concentrate is indicated for the urgent reversal of warfarin-induced coagulopathy in patients with bleeding, but has also been used in an off label use for the treatment of patients with factor X deficiency as well as for the reversal of direct factor X inhibitors in patients with bleeding. The three factor concentrate is indicated for the management of patients with hemophilia B.

Activate prothrombin complex concentrate (aPCC) is a plasma-derived product that contains activated factor VII as well as nonactivated factor II, factor IX, and factor X. This product is derived from heat-treated plasma and indicated for the management of patients with hemophilia A or B with the presence of an inhibitor. Recombinant factor VIIa is a nonplasma-derived form of activated factor VII that is also indicated for the use of patients with hemophilia A or B with the presence of an inhibitor, but also indicated for the use in patients with congenital factor VII deficiency, Glanzmann thrombasthenia, or patients with acquired hemophilia. It is also commonly used in an off label used for severe bleeding such as in trauma, intracranial, postpartum, and reversal of direct oral anticoagulants associated with bleeding.

FIBRIN SEALANT

Fibrin sealant is a product containing fibrinogen and thrombin and when mixed form a fibrin clot. It is typically used for topical application to obtain hemostasis. These products are typically derived from human plasma, however there are bovine preparations available that may contain residual factor V and have been associated with the development of a clinically significant factor V inhibitor.

GENERAL BLOOD PRODUCT ADMINISTRATIVE CONSIDERATIONS

Before blood products can be administered, appropriate consent must be obtained from the recipient with explanation of all risk, benefits, complications, and alternatives to transfusion and verification of consent should be reviewed before administration of any blood components. Additionally, at the time of transfusion, review of the transfusion order should be performed as well as correct identification of the recipient with verification of the blood component for that patient (label identification of patient and unit, blood type, donor identification number, appearance of the unit, and expiration date and time) should be performed at the bedside by two licensed health care professionals.

Blood components can be transfused though catheters ranging in size from 22–14 gauge [32]. Typically, the IV tubing is primed with normal saline (0.9%) or the blood component. There is a mechanism in place, usually with a Y connector, to bypass the blood component and maintain an open venous access should the transfusion need to be stopped. Additionally, as part of the tubing set, all blood products must be transfused though a macroaggregate filter (170–260 micron). This is done to remove and prevent transfusion of any clots or large debris to the patient. If normal saline is not used to maintain the line, other compatible solutions such as ABO compatible-plasma, 5% albumin, PFFs, Normosol-R, and Plasma-Lyte-A can be used as alternative solutions. Medications should never be transfused with blood products. Solutions that are not isotonic, such as solutions containing dextrose/half normal saline, or solutions with high concentration of calcium, such as lactated ringers, are not acceptable as this may lead to hemolysis of RBC products due to osmosis or clotting due to excess chelation of calcium to the citrate anticoagulation, respectively.

At the initiation of the transfusion, baseline vital signs of the patient are obtained and the transfusion should be started at a slow rate, typically 1–2 mL/min to observe the patient for any possible transfusion reaction during the first 15 min. After the initial 15 min, if no transfusion reaction is observed then the rate can be increase. Typically transfusing 2 mL/min for granulocytes, 4 mL/min for RBCs, and 5 mL/min for plasma-based products (platelets, plasma, and cryo) or as tolerated by the patient are standard practice. Whatever rate is selected, transfusion must be completed within 4 h. If this is an issue, such in pediatric patients or patients that are sensitive to volume overload, steps should be taken to split the unit and provide aliquots to allow transfusion of the product in the appropriate timeframe of 4 h. After the transfusion is completed, post vital signs are obtained and the tranfusionist should also document the volume transfused as well as if any transfusion reactions were noted.

Blood warmers can also be used during transfusion to deter hypothermia and are typically used in the setting of rapid transfusion of cold products (RBC/ plasma) such as during trauma, surgery, apheresis, and in newborns. Although there is no clear-cut benefit, blood warms can also be used in patients with cold agglutinin disease [2].

KEY POINTS

- RBC transfusion should be considered in hemodynamically stable hospital patients that have an Hgb <7 g/dL. For surgical patients or those with underlying cardiovascular disease, a Hgb <8 g/dL should prompt consideration of RBC transfusion. Transfusion strategies for certain population, such as those with acute coronary syndrome, lack sufficient evidence to determine appropriate transfusion thresholds.

- One unit of RBCs is expected to increase the Hgb by 1 g/dL or Hct by 3% in an average size adult. In pediatric patients, a typical dose of 10–15 cc/kg is expected to increase the Hgb by 2–3 g/dL or Hct by 6%.
- ABO identical or compatible is acceptable for RBC component therapy but ABO identical should be used for whole blood transfusion due to the increased amount of plasma and therefore isohemagglutinins in whole blood products.
- For all RBC transfusion, confirmation of ABO compatible is performed, usually by immediate spin or electronic crossmatch. If the patient has alloantibodies, a serologic crossmatch to the AHG phase is performed.
- Prophylactic platelet transfuse should be given to patients with platelet counts <10,000/μL due to the risk of spontaneous bleeding.
- One dose of platelets must contain at least 3×10^{11} platelets (either one apheresis unit or approximately 6 units of whole blood derived plates) and is expected to increase the platelet count by 30,000–60,000/μL in an average size adult. Therefore, one unit of whole blood derived platelets would be expected to increase the platelet count by 5000–10,000/μL.
- Platelets contain ABO antigens. As a result, ABO match or compatible units in regards to the platelet antigens (Type O platelet transfuses into a Type A recipient) will be the most efficacious.
- The risk of hemolysis is approximately 1:3000 to 1:10,000 for transfusion of ABO plasma incompatible platelets (Type O platelet transfused into a Type A recipient)
- The CCI is a calculation that can be performed to determine the presence of platelet refractoriness and defined by having a CCI <5000–7500.
- $$CCI = \frac{\text{Body surface area (m}^2) \times [\text{post} - \text{pre transfusion platelet count} \times 10^{11}]}{\text{Number of platelets transfused} \times 10^{11}}$$
- Platelets do not contain Rh antigens but platelet products may contain residual RBCs. Although the risk of alloimmunization is unlikely from Rh positive platelet transfusions into Rh negative individual, Rh negative woman of child bearing age that receive Rh positive platelets should be given RhIg due to the clinical consequence of anti-D alloimmunization during pregnancy.
- Plasma transfusions are indicated for patients with active bleeding in the setting of multifactor deficiency or reversal of warfarin-induced coagulopathy. It is also indicated in patients with factor deficiency in which no factor concrete is available and as a replacement fluid for apheresis such as for patients with TTP.
- Cryo is used primarily for the treatment of fibrinogen deficiency and cannot be used as a low-volume substitute for plasma as it only contains factor VIII, factor XIII, and fibrinogen (factor I), vWF, and fibronectin.
- Granulocyte transfusion should be considered in patients with an ANC <500/μL, bacterial or fungal infection for 24–48 h, unresponsive to antimicrobials, and a chance of marrow recovery.

- Granulocyte should be ABO compatible as a result of the high contamination by RBCs. Additionally, granulocytes should be irradiated to prevent TA-GVHD. CMV seronegative granulocyte units should also be provided to CMV-negative recipients since leukoreduction cannot be performed.

REFERENCES

[1] Blood transfusion therapy: a physicians handbook. American association for Blood Banks (AABB); 2014.

[2] Fung M, Grossman B, Hillyer C, Westhoff C. Technical manual. 18th ed. American association for Blood Banks (AABB); 2014.

[3] Wang JK, Klein HG. Red blood cell transfusion in the treatment and management of anemia: the search for the elusive transfusion trigger. Vox Sang 2010;98:2–11.

[4] Hébert PC, Wells G, Blajchman MA, Marshall J, et al. A multicenter, randomized, controlled clinical trial of transfusion requirements in critical care Transfusion Requirements in Critical Care Investigators, Canadian Critical Care Trials Group. N Engl J Med 1999;340:409.

[5] Carson JL, Guyatt G, Heddle NM, Grossman BJ, et al. Thresholds and storage: clinical practice guidelines from the AABB: red blood cell transfusion. JAMA 2016;316: 2025–35.

[6] Carson JL, Terrin ML, Noveck H, Sanders DW, et al. Liberal or restrictive transfusion in high-risk patients after hip surgery. N Engl J Med 2011;365:2453–62.

[7] Murphy GJ, Pike K, Rogers CA, Wordsworth S, et al. Liberal or restrictive transfusion after cardiac surgery. N Engl J Med 2015;372:997–1008.

[8] Hajjar LA, Vincent JL, Galas FR, Nakamura RE, et al. Transfusion requirements after cardiac surgery: the TRACS randomized controlled trial. JAMA 2010;304:1559–67.

[9] Bracey AW, Radovancevic R, Riggs SA, Houston S, et al. Lowering the hemoglobin threshold for transfusion in coronary artery bypass procedures: effect on patient outcome. Transfusion 1999;39:1070–7.

[10] Kaufman RM, Djulbegovic B, Gernsheimer T, Kleinman S, et al. Platelet transfusion: a clinical practice guideline from the AABB. Ann Intern Med 2015;162:205–13.

[11] Gmür J, Burger J, Schanz U, Fehr J, et al. Safety of stringent prophylactic platelet transfusion policy for patients with acute leukemia. Lancet 1991;338(8777):1223–6.

[12] Bishop JF, Schiffer CA, Aisner J, Matthews JP, et al. Surgery in acute leukemia: a review of 167 operations in thrombocytopenic patients. Am J Hematol 1987;26:147–55.

[13] Hartman M, Sucker C, Boehm O, Koch A, et al. Effects of cardiac surgery on hemostasis. Transfus Med Rev 2006;20:230–41.

[14] Mannucci PM, Remuzzi G, Pusineri F, Lombardi R, et al. Deamino-8-D-arginine vasopressin shortens the bleeding time in uremia. N Engl J Med 1983;308(1):8–12.

[15] Sandler SG. Treating immune thrombocytopenic purpura and preventing Rh Alloimmunization using intravenous rho (d) immune globulin. Transfus Med Rev 2001;15:67–76.

[16] Otrock ZK, Liu C, Grossman BJ. Platelet transfusion in thrombotic thrombocytopenic purpura. Vox Sang 2015;109:168–72.

[17] Zhou A, Mehta RS, Smith RE. Outcomes of platelet transfusion in patients with thrombotic thrombocytopenic purpura: a retrospective case series study. Ann Hematol 2015;94:462–72.

[18] Kumar R, Mehta RS, Zhou A, Smith RE. Outcomes of platelet transfusion in heparin induced thrombocytopenia patients. Blood 2013;. 122:2311(ASH Abstract).

[19] Vassallo RR, Murphy S. A critical comparison of platelet preparation methods. Curr Opin Hematol 2006;13:323–30.

[20] Shehata N, Tinmouth A, Naglie G, Freedman J, et al. ABO-identical versus nonidentical platelet transfusion: a systematic review. Transfusion 2009;49:2442–53.

[21] Larson LG, Welsh VJ, Ladd DJ. Acute intravascular hemolysis secondary to out of group platelet transfusion. Transfusion 2000;40:902–6.

[22] Mair B, Benson K. Evaluation of changes in hemoglobin levels associated with ABO-incompatible plasma in apheresis platelets. Transfusion 1998;38:51–5.

[23] Carr R, Hutton JL, Jenkins JA, Lucas GF. Transfusion of ABO-mismatched platelets leads to early platelet refractoriness. Br J Haematol 1990;75:408–13.

[24] Cid J, Carvasse G, Pereira A, Sanz C, et al. Platelet transfusion from D+ donors to D- patients: a 10 year follow up study of 1014 patients. Transfusion 2011;51:1163–9.

[25] Szczepiorkowski ZM, Dunbar NM. Transfusion guidelines: when to transfuse. Hematol Am Soc Hematol Educ Program 2013;2013:638–44.

[26] Segal JB, Dzik WH. Paucity of studies to support that abnormal coagulation test results predict bleeding in the setting of invasive procedures: an evidence-based review. Transfusion 2005;45:1413–25.

[27] Burns ER, Goldberg SN, Wenz B. Paradoxic effect of multiple mild coagulation factor deficiencies on the prothrombin time and activated partial thromboplastin time. Am J Clin Pathol 1993;100:94–8.

[28] Holland LL, Foster TM, Marlar RA, Brooks JP. Fresh frozen plasma is ineffective for correcting minimally elevated international normalized ratios. Transfusion 2005;45:1234–5.

[29] Janson PA, Jubelirer SJ, Weinstein MJ, Deykin D. Treatment of the bleeding tendency in uremia with cryoprecipitate. N Engl J Med 1980;303:1318–22.

[30] Remuzzi G. Bleeding in renal failure. Lancet 1988;1(8596):1205–8.

[31] Klein K, Castillo B, Historical Perspectives. Current status, and ethical issues in granulocyte transfusion. Ann Clin Lab Sci 2017;47:5010–507.

[32] De la Roche MR, Gauthier L. Rapid transfusion of packed red blood cells: Effects of dilution, pressure, and catheter size. Ann Emerg Med 1993;22:1551–5.

Patient blood management and utilization

INTRODUCTION

Patient blood management is an evidence-based, multidisciplinary approach to optimizing care of patients who might require transfusion. It encompasses all aspects of patient evaluation and clinical management surrounding the transfusion decision-making process, including the application of appropriate indications, as well as minimization of blood loss and optimization of patient red cell mass [1]. Important points regarding blood transfusion include:

- Blood transfusion is the most common procedure performed during hospitalization.
- 11% of all hospital stays with a procedure include transfusion.
- 50% of red blood cell (RBC) transfusions are found inappropriate.
- Nearly 14 million allogeneic red cell units are transfused each year at a cost to hospitals of over $ 3 billion (average red cell $ 225/unit).
- Activity-based cost analysis shows that the cost of transfusion independent of complications associated with the transfusion range from $ 800 per unit to over $ 1200 per transfused unit.

ANEMIA MANAGEMENT

Anemia and iron deficiency are extremely common and widely underrecognized condition. It is estimated that greater than one-third of adults over the age of 65 have unexplained anemia, defined as hemoglobin (Hb) less than 12 g/dL. Seventeen percent of adults over the age of 65 have been shown to have iron deficiency and of those with iron deficiency anemia, only 50% normalized their Hb with oral iron therapy. Iron deficiency in the hospitalized patient population, both functional and true iron deficiency, is also common [2,3].

Hospital acquired anemia is prevalent which has a major health care implications because such anemia has significant impact on clinical outcome. Preoperative anemia is the most important predictor of perioperative transfusion [4,5].The association of preoperative anemia with postoperative mortality is a compelling reason to manage anemia in the perisurgical patient population.

Table 8.1 Intravenous Iron Formulations

Iron Formulation	Brand Name	Concentration of Elemental Iron	Test Dose	Dosing (Adults)
Sodium ferric gluconate	Ferrlecit	12.5 mg/mL	No	Multiple doses of 125–187.5 mg
Ferumoxytol	Feraheme	30 mg/mL	No	Two doses of 510 mg given 3–8 days apart, or Single dose of 1020 mg
Iron dextran	INFeD, CosmoFer	50 mg/mL	Yes, 25 mg (0.5 mL) prior to the first dose	Multiple doses of 100 mg, or Single dose of 1000 mg diluted in 250 mL normal saline
Iron sucrose	Venofer	20 mg/mL	No	Multiple doses of 200–300 mg

Anemia recognition, diagnosis, and treatment will reduce the cost and complications of transfusion. Enteric iron therapy is ineffective and may be harmful. Side effects associated with oral iron are well known. These complications often lead to noncompliance because as many as 40%–50% of patients are unable to tolerate enteric iron therapy. In patients treated with an erythropoiesis stimulating agent (ESA), intravenous iron is much more efficacious than enteric iron in ensuring an adequate response to the ESA at the lowest dose. Various intravenous iron preparations available for treating anemia are listed in Table 8.1.

INFORMED CONSENT

Patient blood management program should be participating in the development and revision of policies, processes, and procedures regarding patient consent for transfusion and the right to decline transfusion. The patient should be provided with resources alternative to blood transfusion. At a minimum, elements of consent shall include all of the following:

1. A description of the risks, benefits, and treatment alternatives.
2. The opportunity to ask and receive answers to the questions.
3. The right to accept or refuse treatment [6].

COAGULATION MANAGEMENT

Taking medical history is the most efficient approach to evaluate bleeding risk of a patient. Spontaneous bruising and bleeding, excessive bleeding with stress (menses, dental extraction, and surgery), and medication use (dose, timing of the last dose, mechanism, and pharmacokinetics are important clinical consideration) are very useful information to determine risk of bleeding in a patient. Conventional blood tests, platelet count, prothrombin time (PT), and activated partial thromboplastin time (aPTT) rarely detect the drug effect and have very limited utility in the preoperative setting except for patients with thrombocytopenia, hemophilia, and liver dysfunction. Specific tests for antiplatelet drugs like the Verify Now assays may also be inaccurate. It is very important in the preoperative stage to correct anemia if present in the patient as well as to be aware of any known coagulation problem in the patient. It is also important that the patient has adequate liver function. During the intraoperative period, it is important to avoid hypothermia, acidosis, hypocalcemia, and other conditions that may result in adverse clinical outcome.

It is preferable to stop therapy with anticoagulant prior to surgery rather than using reversing agents to counteract effects of anticoagulants. Patients who had history of heart disease with stents or uncorrected vascular lesions, cerebrovascular vascular stents, abdominal aortic aneurysms with prostheses, and celiac and portal vascular thrombosis might be on antiplatelet or anticoagulant medications. Various antiplatelet as well as anticoagulant agents are used for such therapy and there are recommendations when such therapy must be stopped prior to surgery [7–9]. Various aspects of these medications are listed in Table 8.2.

INDICATION FOR TRANSFUSION

According to the 2011 National Blood Collection and Utilization survey report, 92% of survey respondents reported the use of transfusion guidelines. While many institutions have institution-specific guidelines, 85% of the guidelines used were predominantly based on one of the national guidelines. (American Association for Blood banks 50%, College of American Pathologists 25.3%, American Red Cross 8.6%, American Society of Anesthesiologists 1.3%, other 14.8%). Other hospitals indicated that they based their transfusion guidelines on recommendations from The Joint Commission, the New York State Department of Health, the American Society of Hematology, the hospital's own internal transfusion committee, and/or multiple sources of evidence-based practices.

There are eight landmark randomized clinical trials that support Hb level between 7 and 8 g/dL as indication for transfusion as a part of restrictive strategy [10–17]. Key findings from these trails are summarized in Table 8.3.

INTRAOPERATIVE TRANSFUSION STRATEGIES AND SURGICAL TECHNIQUES

There are great opportunities for better utilization of blood products by reducing use of such products at the intraoperative setting. Applying multiple strategies can be vital for successful blood less surgery. Numerous techniques and devices are available for reducing the loss of blood during surgical procedures. There are multiple topical hemostatic agents available to reduce localized blood loss.

Table 8.2 Mechanism of Action and Guidelines for Stopping Therapy With Antithrombotic Agents Prior to Surgery

Agent	Mechanism of Action	Recommended Interval Between Last Dose and Surgery
Antiplatelet agents		
Aspirin	Irreversible cyclooxygenase inhibitor	7–10 days
Aspirin and dipyridamole	Phosphodiesterase inhibitor	7–10 days
Cilostazol	Phosphodiesterase inhibitor	7–10 days
Thienopyridine agents	ADP receptor antagonist	
Clopidogrel		5 days (clopidogrel and ticagrelor)
Ticlopidine		
Prasugrel		7 days (prasugrel)
Ticagrelor		10–14 days (ticlopidine)
Anticoagulant agents		
Warfarin	Inhibition of vitamin K dependent factors for γ carboxylation and protein C and S	1–8 days, depending on INR and patient characteristics: INR decreases to ≤ 1.5 in approximately 93% of patients within 5 days
Unfractionated heparin	Antithrombin activation (inhibition of factors IIa, IXa, IXa, and XIIa)	IV, 2–6 h depending on dose. Subcutaneous, 12–24 h depending on dose
Low-molecular-weight heparin (LMWH)	Antithrombin activation (inhibition of factors IIa, IXa, XIa, and XIIa)	24 h
Fondaparinux	Antithrombin activation (factor Xa inhibitor)	36–24 h
Bivalirudin	Direct thrombin inhibitor	2–6 h, depending on dose and creatinine clearance
Dabigatran	Direct thrombin inhibitor	1 or 2 days with creatinine clearance rate of ≥50 mL/min; 3–5 days with creatinine clearance rate of <50 mL/min
Rivaroxaban	Direct factor Xa inhibitor	≥1 day with normal renal function; 2–4 days with creatinine clearance rate of <90 mL/min
Apixaban	Direct factor Xa inhibitor	1 or 2 days with creatinine clearance of >60 mL/min; 3–5 days with creatinine clearance of <60 mL/min
Desirudin	Direct thrombin inhibitor	2 hr.
Edoxaban	Factor Xa inhibitor	24 h
Betrixaban	Factor Xa inhibitor	24–48 h

Table 8.3 Hemoglobin Levels Triggering Transfusion as Reported by Eight Landmark Clinical Trails

Patient Population Studied	Hemoglobin Trigger for Restrictive Strategy	Comments	Year Published	References
Medical intensive care unit patients; Canadian Critical Care Trial Group	7 g/dL	Restrictive strategy is as effective as and possibly superior to a liberal strategy (Hb ≤ 10 g/dL) for RBC transfusion	1999	[10]
Pediatric intensive care unit patients; Canadian Critical Care Trial Group and Pediatric Acute Lung Injury and Sepsis Investigators Net Work	7 g/dL	Restrictive strategy and liberal strategy (Hb ≤ 9.5 g/dL) showed similar outcome for RBC transfusion	2007	[11]
Cardiac surgery patients; TRACS (Transfusion Requirement After Cardiac Surgery) trial	Hematocrit (Hct) 24%	Restrictive strategy (Hct 24%) and liberal strategy (Hct: 30%) showed similar outcome for RBC transfusion	2010	[12]
Elderly (age 50 years or older) high risk orthopedic surgery patients	8 g/dL	Restrictive strategy and liberal strategy (Hb ≤ 10 g/dL) showed similar outcome for RBC transfusion	2011	[13]
Patients with acute upper gastrointestinal bleeding	7 g/dL	Restrictive strategy produced superior patient outcome than liberal strategy (Hb ≤ 9 g/dL) for RBC transfusion	2013	[14]
Patients with septic shock	7 g/dL	Restrictive strategy and liberal strategy (Hb ≤ 9 g/dL) showed similar outcome for RBC transfusion	2014	[15]
Patients with traumatic brain injury	7 g/dL	Restrictive strategy showed similar possibly better outcome than liberal strategy (Hb ≤ 10 g/dL) showed similar outcome	2014	[16]
Cardiac Surgery patients	7.5 g/dL	Restrictive strategy and liberal strategy (Hb < 9 g/dL) showed similar outcome for RBC transfusion	2015	[17]

In some major surgical procedures, significant blood loss is inevitable. Using alternatives to red cell transfusion can improve safety of blood transfusion. Historically, preoperative autologous donation (PAD) was used. Intraoperative blood loss is unpredictable and half of the PAD units are typically discarded. As a result, many hospitals discourage PAD technique [18].

Acute normovolemic hemodilution (ANC) involves removal of whole blood just before the start of surgery so that bloodshed during surgery has a lower concentration of red cells. The harvested whole blood is then returned upon completion of surgery. ANC can limit red cell use. The recommend degree of hemodilution is controversial. Some physicians remove blood to a Hb level of 6 g/dL, whereas others are concerned at risk for ischemic events and limit their dilution to much higher Hb concentrations.

Blood recovery or cell salvage is the collection and reinfusion of shed blood. Shed blood can be washed and filtered or simply filtered. Intraoperative blood recovery (washed and filtered) can be used for cardiothoracic, orthopedic, neurosurgery, obstetrics, gynecology, urology, and vascular procedures. Unwashed recovered blood is usually reserved for the postoperative environment (e.g., ICU), where small quantities of blood are collected and reinfused. Salvage cells must be reinfused within 6 h of the commencement of collection, in order to minimize the risk of infection.

ICU STRATEGIES AND ANCILLARY TECHNIQUES

Studies in multidisciplinary ICU indicate that an average of 40–70 mL/day of blood is drawn solely for diagnostic purpose. For each 100 mL of blood collected for laboratory testing, a patient's Hb level may decrease an average of 0.7 g/dL. Phlebotomy blood loss may account for as much as 30% of blood transfused in the ICU [19]. In an effort to reduce blood loss due to diagnostic test, low volume adult blood sampling tubes may be used. In addition, blood conserving arterial/central venous line systems may also be utilized for this purpose.

Point of-care testing can be utilized easily in hospital setting such as ICU, emergency room, etc. when rapid access to laboratory results are needed. Currently various point of care testing devices are readily available for determination of Hb/hematocrit, INR (international normalized ratio), platelet function and whole blood coagulation monitoring such as thromboelastography (TEG) and thromboelastometry (also known as rotational thromboelastometry: ROTEM). Point of care testing devices allow for smaller blood volumes and can prevent hospital acquired anemia. Quick results also enable instant interventions in critical situations as goal directed care.

BLOOD UTILIZATION AND REVIEW

Blood utilization review is a required and necessary function for all hospital transfusion services. The Joint Commission requires hospitals to collect data to monitor the performance of proce sses that involve risks or may result in sentinel events, which includes the use of blood and blood components. The American Association for Blood Banks (AABB) requires all facilities to have a peer-review program to monitor transfusion practices for all categories of blood and components, including monitoring of usage and discard and appropriateness of use. Medicare and Medicaid require that hospitals make recommendations to the medical staff regarding improvements in transfusion procedures.

There are three types of blood utilization review:

1. Prospective (real-time) review
2. Concurrent review
3. Retrospective review

PROSPECTIVE (REAL-TIME) REVIEW

In general, transfusion requests are reviewed in real time (i.e., review of individual transfusion requests before issue of the components). This review provides an opportunity for the medical professional in the blood bank to intervene or even stop or modify an inappropriate transfusion requests. Individual transfusion requests are reviewed using prespecified guidelines.

CONCURRENT REVIEW

Review of individual transfusion requests that occur in the 12–24 h following the transfusion episode should also be instituted. As a post transfusion review, the individual transfusion event cannot be altered. The concurrent review can only alter the future blood transfusion practice.

RETROSPECTIVE REVIEW

Retrospective review offers the opportunity to review aggregate transfusion data and trend in transfusion utilization. The mean or median number of components units transfused per hospitalized patient and/or procedures provides a more meaningful summary of component use. Result should be reviewed by the transfusion medicine/patient blood management committee.

All forms of review can be used to effect changes in transfusion practice. Utilization reviews may be combined with other interventions such as educations, practical usage of guidelines, and the attending physician accountable.

THE JOINT COMMISSION PATIENT BLOOD MANAGEMENT CERTIFICATION

The Joint Commission and AABB have teamed up to offer Patient Blood Management Certification. This new voluntary certification is available to hospitals which are accredited by The Joint Commission [1]. Process to certification is based on the AABB guidelines for Standards for a Patient Blood Management Program. Implementing patient blood management program has following benefits:

- Risk reduction resulting in fewer adverse events and incidents
- Improved patient outcomes
- Reduced hospital stays, readmissions, and length of stay
- Ensuring blood availability for those most in need
- Optimized care for those who may need transfusion
- Fostering collaboration throughout the hospital
- Providing a competitive edge in the marketplace
- Cost saving

KEY POINTS

- Patient blood management is an evidence-based, multidisciplinary approach to optimizing care of patients who might need transfusion. It encompasses all aspects of patient evaluation and clinical management surrounding the transfusion decision-making process, including the application of appropriate indications, as well as minimization of blood loss and optimization of patient red cell mass.
- It is estimated that greater than one-third of adults over the age of 65 have unexplained anemia, defined as Hb less than 12 g/dL. Seventeen percent (17%) of adults over the age of 65 have been shown to have iron deficiency and of those with iron deficiency anemia, only 50% normalized their Hb with oral iron therapy.
- The association of preoperative anemia with postoperative mortality is a compelling reason to manage anemia in the perisurgical patient population.
- In patients treated with an ESA, intravenous iron is much more efficacious than enteric iron.
- Taking detail medical history is very effective in evaluating bleeding risk of a patient scheduled for surgery. It is very important to maximize patient condition from the preoperative period by minimizing anemia, correcting or plan for known coagulation problems, and maximizing hepatic function.
- There are eight landmark randomized clinical trials supporting Hb Triggers of 7–8 g/dL as indication of RBC transfusion.
- Many hospitals do not recommend PAD. Blood recovery and ANC are common approaches used in perioperative patient care.
- Studies in multidisciplinary ICU patients show that an average of 40–70 mL/day of blood is drawn solely for diagnostic purpose. For each 100 mL of blood collected for laboratory testing, a patient's Hb level may decrease an average of 0.7 g/dL. Phlebotomy blood loss may account for as much as 30% of blood transfused in the ICU.

- Point-of-care testing devices are available in ICU, emergency room, etc,. for bedside testing of blood to determine Hb/hematocrit, INR, and platelet function. In addition, whole blood coagulation monitoring (TEG and ROTEM) is also possible using point of care devices.
- Blood utilization review is a required and necessary function for all hospital transfusion services.
- There are three types of blood utilization review, prospective (real-time) review, concurrent review, and retrospective review.
- The Joint Commission and AABB have teamed up to offer Patient Blood Management Certification. This new voluntary certification is available to hospitals and critical access hospitals that are accredited by The Joint Commission.

REFERENCES

[1] The Joint Commission: Patient Blood Management Certification. Available from: https://www.jointcommission.org/certification/patient_blood_management_certification.aspx.
[2] Walsh TS, Saleh EE. Anemia during critical illness. Br J Anaesth 2006;97:278–91.
[3] Rodriguez RM, Corwin HL, Gettinger A, Corwin MJ, et al. Nutritional deficiencies and blunted erythropoietin response as causes of the anemia of critical illness. J Crit Care 2001;16:36–41.
[4] Millett PJ, Porramatikul M, Chen N, et al. Analysis of transfusion predictors in shoulder arthroplasty. J Bone Joint Surg Am 2006;88:1223–30.
[5] Faris PM, Spence RK, Larholt KM, et al. The predictive power of baseline hemoglobin for transfusion risk in surgery patients. Orthopedics 1999;22(Suppl. 1):s135–40.
[6] Standards for a Patient Blood Management Program. 1st ed. American Association of Blood Banks (AABB); Bethesda, MD 1st ed.; 2014.
[7] Baron TH, Kamath PS, McBane RD. Management of antithrombotic therapy in patients undergoing invasive procedures. N Engl J Med 2013;368:2113–24.
[8] Bracey AW, Reyes MA, Chen AJ, Bayat M, et al. How do we manage patients treated with antithrombotic therapy in the preoperative interval. Transfusion 2011;51:2066–77.
[9] Angiolillo DJ, Firstenberg MS, Price MJ, Tummala PE, et al. Bridging anti-platelet therapy with the intravenous agent cangrelor in patients undergoing cardiac surgery. JAMA 2012;307:265–74.
[10] Hebert PC, Wells G, Blajchman MA, Marshall J, et al. A multicenter, randomized, controlled clinical trial of transfusion requirements in critical care investigators. Can Crit Care Trial Group 1999;340:409–17.
[11] Lacroix J, Hebert PC, Hutchison JS, Hume HA, et al. Transfusion strategies for patients in pediatric intensive care units. N Engl J Med 2007;356:1609–19.
[12] Hajjar LA, Vincent JL, Galas FR, Nakamura RE, et al. Transfusion requirements after cardiac surgery: the TRACS randomized controlled trial. JAMA 2010;304:1559–667.
[13] Carson JL, Terrin ML, Noveck H, Sanders H, et al. Liberal or restrictive transfusion in high risk patients after hip surgery. N Engl J Med 2011;365:2453–62.
[14] Villanueva C, Colombo A, Bosch A, Concepcion M, et al. Transfusion strategies for acute upper gastrointestinal bleeding. N Engl J Med 2013;368:11–21.

[15] Holst LB, Hasse N, Wetterslev J, Wernerman J, et al. Lower versus higher hemoglobin threshold for transfusion in septic shock. N Engl J Med 2014;371:1381–91.

[16] Robertson GS, Hannay HJ, Yamal JM, Gopinath S, et al. Effect of erythropoietin and transfusion threshold on neurological recovery after traumatic brain injury: a randomized clinical trial. JAMA 2014;312:36–47.

[17] Murphy GJ, Pike K, Rogers CA, Wordsworth S, et al. Liberal or restrictive transfusion after cardiac surgery. N Engl J Med 2015;372:997–1008.

[18] Brecher ME, Goodnough LT. The rise and fall of preoperative autologous blood donation. Transfusion 2001;41:1459–62.

[19] Napolitano LM, Kurek S, Luchette FA, Anderson GL, et al. Clinical practice guideline: red blood cell transfusion in adult trauma and critical care. J Trauma 2009;67: 1439–42.

Introduction to quality management system

INTRODUCTION

Proper quality control (QC) and quality management is essential for running a successful blood bank service. A quality management system includes the organizational structure, responsibilities, polices, processes, procedures, and resources established by executive management to achieve and maintain quality.

The establishment of a formal quality assurance (QA) program is required as per policies of Centers for Medicare and Medicaid Services (CMS), Clinical Laboratory Improvement Amendments (CLIA) [1], and the Food and Drug Administration (FDA). Moreover, the current good manufacturing practice (cGMP) and current good tissue practice (cGTP) regulations are also applicable in certain aspects of quality management of blood bank service [2–5].

The FDA regulations in the Code of Federal Regulations (CFR) Title 21, Part 211.22 require an independent QC or QA unit that has responsibility for the overall quality of the facility's finished product and authority to control the processes that may affect this product [3]. Professional and accrediting organizations such as the American Association for Blood Banks (AABB), College of American Pathologist (CAP), the Joint Commission, Clinical and Laboratory Standards Institute (CLSI), and Foundation of the Accreditation of Cellular Therapy (FACT), have also established requirements and guidelines to address quality issues in clinical laboratories.

The Internal Organization for Standardization (ISO) quality management standards (ISO 9001) are generic to any industry and describe the important minimum elements of a quality management system [6]. The ISO 15189 standards are specific to laboratory medicine. The AABB Quality System Essentials (QSEs) were developed to be compatible with ISO 9001 standards, the FDA Guideline for Quality Assurance in Blood Establishments, and other FDA Quality system approaches.

QUALITY CONTROL

Quality control (QC) in clinical laboratory, including blood bank, is essential to produce accurate result for laboratory test results. QC provides opportunity to laboratory professionals to detect and correct deficiencies in a laboratory's internal analytical process prior to the release of patient results in order to minimize laboratory errors.

Transfusion Medicine for Pathologists. http://dx.doi.org/10.1016/B978-0-12-814313-1.00009-5

QC accesses the suitability of inputs to determine whether the output meets specifications. Output may be products or services. Historically, transfusion services and donor centers have used many QC measures as standard practices in their operations. The following are examples of QC functions: reagent QC, product QC, clerical checks, visual inspections, and measurements, such as temperature readings on refrigerators and volume or cell counts on finished blood components. The frequency for QC testing is determined by the facility in accordance with the applicable CMS, FDA, AABB, State and manufacturer requirements. Unacceptable QC results must be investigated and corrective action must be implemented, if indicated, before the QC procedure is repeated or the operational process is continued.

QUALITY ASSURANCES

Quality assurances (QA) activities are not tied to the actual performances of a QC step. The purpose of QA is to control processes, detect unwanted shifts or trends that may require correction, and identify opportunities for process improvement. QA activities include reviewing procedures and policies for compliance with regulatory and accrediting agency guidelines, reviewing operational performance data, reviewing records for completeness and accuracy, performing audits, monitoring quality indicators, and internal assessments.

QUALITY MANAGEMENT

Quality management considers interrelated processes in the context of the organization and its relations with customers and suppliers. It addresses the leadership role of executive management in creating a commitment to quality throughout the organization, the understanding of suppliers and customers as partners in quality, the management of human and other resources, and quality planning.

REGULATIONS IN CLINICAL LABORATORIES

Regulations are the rules under which the requirements of law are enforced. Federal agencies such as the FDA and the CMS are regulatory agencies which have authority to oversight operation of clinical laboratories.

The FDA regulates blood establishments under the authority of the Food, Drug, and Cosmetics Act of 1938 (under which blood components are considered drugs) and the Public Health Service Act of 1944. Applicable regulation are listed in the CFR, Title 21, Parts 200–299 (labeling, cGMP), 600–680 (biological products), and 800–898 (medical devices, adverse events). Human cells, tissues, and cellular as well as tissue-based products (HCT/Ps) are regulated under The Public Health Service Act of 1944, Section 361. Applicable regulations are found in 21 CFR Parts 1270–1271. HCT/Ps mean articles containing or consisting of human cells or tissues that are intended for implantation, transplantation, infusion, or transfer into a human recipient.

CMS regulates US medical laboratories under the authority of the Public Health Service Act of 1944, Section 353, and the Clinical Laboratory Improvement Amendments of 1988 (CLIA'88). CLIA regulations are found in 42 CFR Part 493. In addition to federal regulatory agencies, there may be state and local-level regulatory agencies that have jurisdiction over blood and tissue establishments. Accreditation is certification that is issued by an appropriate regulatory agency when an establishment has met the quality standards set forth by an agency. Inspection of a laboratory for the purpose of accreditation or renewal of accreditation may be performed by peers, for example inspection team for CAP accreditation usually includes pathologists and laboratory professionals.

The AABB also has accreditation program for blood banks which promotes the highest standards of care for both patients and donors covering all aspects of blood banking, transfusion medicine, relationship testing, hematopoietic, cord blood and other cellular therapies. In May 2014, AABB was granted "deemed status" as an accrediting organization under the CLIA'88. Most recently, the accreditation program is accredited by the International Society for Quality in Healthcare (ISQua).

Good manufacturing practice (cGMP) regulations, cGTP regulations, CLIA'88 requirements, increased regulations from the FDA for unlicensed blood establishments, as well as transfusion services, all have positive impact in improving quality of blood bank services as well as reducing risk of transfusion related adverse effects. In addition, the introduction of quality management systems, competition within the blood banking profession, advances in the cellular therapy arena, and increased pressure to improve safety with limited resources have changed the business model used by blood banks. Accreditation from AABB ensures that the blood bank is working efficiently and also in compliance with AABB Standards and federal regulations. Verification of compliance is accomplished by peer review assessments performed in a competent and reliable manner. This includes an audit of the quality and operational systems. The experience and expertise of the AABB assessors also makes the assessment process an educational experience [7].

JURAN'S QUALITY TRILOGY

There are three fundamental processes for managing quality in any organization: planning, control, and improvement [8].

During the planning phase, the following steps must be performed:

1. Establish quality goals for the project.
2. Identify the customers.
3. Determine customer needs and expectations.
4. Develop product and service specifications to meet customer, operational, regulatory, and accreditation requirements.

5. Develop operational processes for production and delivery, including written procedures and resources requirements.
6. Develop process controls and validate the process in the operational setting.

Once the plan is implemented, the control process provides a feedback loop for operations that includes the following:

1. Evaluation of performance.
2. Comparison of performance to goals.
3. Action to correct any discrepancy between the two.

Quality improvement is intended to enable an organization to attain higher levels of performance by creating new or better features that add value or by removing deficiencies in the process, product, or service. Improvements must be based on data-driven analysis; an ongoing measurement and assessment program is fundamental to that process.

PROCESS APPROACH

A process includes all the resources and activities that transform an input into an output.
   INPUT → PROCESS → OUTPUT
   For example, a key process for donor centers is donor selection.
   The input includes the individual who presents for donations and all of the resources required for that donor's health screening. Through a series of activities (a process), including the verification of the donor's identity, a deferral status review, a mini-physical exam, and a health history questionnaire, an individual is deemed an eligible donor. The output is either an eligible donor who can continue to the next process (blood collection) or an ineligible donor who is deferred.
   Supplier qualification, formal agreements, supply verification, and inventory control are strategies for ensuring that the inputs to a process meet specifications. Personal training and competence assessment, equipment maintenance and control, management of documents and records, and implementation of appropriate in-process controls provide assurance that the process will operate as intended.
   End-product testing and inspection, customer feedback, and outcome measurement provide data to evaluate product quality and improve the process. These output measurements and quality indicators are used to evaluate the effectiveness of the process and process controls.

APPLICATION OF QUALITY MANAGEMENT PRINCIPLES

Quality management principles include:

1. Organization and leadership
2. Personal

3. Equipment management
4. Purchasing and inventory
5. Process management
6. Information management
7. Documents and records
8. Management and nonconforming events
9. Monitoring and assessment
10. Process improvement
11. Customer focus
12. Facilities, work environment, and safety

ORGANIZATION AND LEADERSHIP

The structure of the organization must be documented, and the roles and responsibilities for the provision of tests, products, and services must be clearly defined. The facility should define in writing the authority and responsibilities of executive management to establish and maintain the quality management system.

The individual designated to oversee the facility's quality functions should report directly to executive management. In addition to having the responsibility to coordinate, monitor, and facilitate quality system activities, this person should have the authority to recommend and initiate corrective action when appropriate [5]. Ideally, this person should be independent of the facility's operational functions. Individuals with dual quality and operational responsibilities should not provide quality oversight for operational work that they have performed.

PERSONAL

The most important laboratory resource is a competent, trained, and motivated staff. Continuous education opportunities should be offered to the staff and recorded. Regular competency assessment and proficiency testing should be conducted and documented.

EQUIPMENT MANAGEMENT

Many kinds of equipment are used in the laboratory, and each piece of equipment must be functioning properly. The steps in equipment management are: choosing the right equipment, installing it correctly, assuring that the staff is properly trained to use the equipment, ensuring the new equipment works properly, and having a system for maintenance.

PURCHASING AND INVENTORY

Proper management of purchasing and inventory can produce cost saving in addition to ensuring supplies and reagents are available when needed. The procedure should be written and implemented to assure that all reagents and supplies are correctly selected, purchased, used, and stored in a manner that preserves integrity and reliability [9].

PROCESS MANAGEMENT

Process management involves managing all activities involved in the operation of a laboratory, from pre-examination to post-examination steps. Process management is comprised of several factors that are important in ensuring the quality of the laboratory testing processes. These factors include QC for testing, appropriate management of the sample, including collection and handling, and method of verification and validation [10].

INFORMATION MANAGEMENT

The FDA considers computerized systems to include hardware, software, peripheral devices, networks, personnel, and documentation [11]. End-user validations of computer systems and the interfaces between systems should be conducted in the environment in which they will be used. Testing by the computer software vendor or supplier is not a substitute for computer validation at the facility. If changes to the computer system result in changes to the process, process revalidation should be performed.

DOCUMENTS AND RECORDS

Documents are written policies, procedures, work instructions, process descriptions, labels and forms. They provide the description how things should happen. Records provide the evidence that what should have performed has performed. When data are recorded in the forms, forms become records. Documents and records may be in paper or electronic form.

There should be defined process to identify, approve, review, revise and retired records and SOPs (standard operating procedures). Documents must be reviewed, modified, and reapproved periodically. All documents are reviewed every 12 months. Documents must be protected from unintended alterations or destruction.

MANAGEMENT AND NONCONFIRMING EVENTS

Quality management system should include a process for detecting, investigating, and responding to events which are deviated from policies, processes, and procedures or failure to meet requirement by facility, AABB standards, or regulations [2–4].

MONITORING AND ASSESSMENT

Assessments can be internal or external, quality assessments, peer reviews, self-assessments, and proficiency testing. There will be comparisons of actual to expected results.

PROCESS IMPROVEMENT

The primary goal of a quality management system is continuous improvement of the laboratory processes, and this must be done in a systemic manner. There are number of tools such as customer service surveys, QC, QA, auditing, and quality indicators, are useful for the process improvement [9].

FOCUS ON CUSTOMERS

A primary focus for any organization interested in quality is serving the needs of its customers. The most appropriate way to ensure the customer needs and expectations are made is to have an agreement, contract, or another document between customers and the facility to ensure customer satisfaction can be achieved. The facility must have a process to address needs and expectations that are not met. For example, for a blood center that has agreed to deliver leukocyte-reduced components daily to customers, it is important for the facility to have adequate processing components in place to supply required high quality leukocyte-reduced components to customers.

Once agreements have been made between the facility and its customers, there should be a means to obtain feedback from the customer to ensure that the facility is meeting the customer's expectations. Any deviation from cGMP that affects the safety, purity, or potency of a distributed blood component, derivative or tissue must be reported to the FDA [2,3].

FACILITIES, WORK ENVIRONMENT, AND SAFETY

The facility should provide a safe workplace with adequate environmental controls and emergency procedures to ensure the safety of patients, donors, staff, and visitors [9,10].

It is important to note that cGMP regulations require quality planning and control of the physical work environment.

KEY POINTS

- Quality Control (QC) is performed for the purpose of evaluating a process in progress. QC accesses the suitability of inputs to determine whether the output meets specifications. Output may be products or services.
- Quality Assurance (QA) activities are not tied to the actual performances of a process. The purpose of QA is to control processes, detect unwanted shifts or trends that may require correction, and identify opportunities for process improvement.
- Quality management considers interrelated processes in the context of the organization and its relations with customers and suppliers.
- Regulations are the rules under which the requirements of law are enforced. Federal agencies such as the FDA and the CMS are regulatory agencies which have also control over operation of clinical laboratories including blood bank.

- Juran's Quality Trilogy is one example of a quality management approach. There are three fundamental processes for managing quality in any organization: planning, control, and improvement.
- A process includes all the resources and activities that transform an input into an output. INPUT → PROCESS → OUTPUT
- Practical Application of quality management principles includes organization and leadership, personal, equipment management, purchasing and inventory, process management, information management, documents and records, management and nonconforming events, monitoring and assessment process improvement, customer focus, facilities, work environment, and safety.

REFERENCES

[1] Code of federal regulations. Title 42, CFR Part 493. Washington, DC: US Government Printing Office, 2013 (revised annually).

[2] Code of federal regulations. Title 21, CFR Parts 606,610,630, and 640. Washington, DC: US Government Printing Office, 2014 (revised annually).

[3] Code of federal regulations. Title 21, CFR Parts 210 and 211. Washington, DC: US Government Printing Office, 2014 (revised annually).

[4] Code of federal regulations. Title 21, CFR Parts 1270 and 1271. Washington, DC: US Government Printing Office, 2014 (revised annually).

[5] Food and Drug Administration. Guideline for quality assurance in blood establishments. Silver Spring, MD: CBER Office of Communication, Outreach, and Development; 1995. (July 11, 1995).

[6] ANSI/ISO/ASQ Q9001-2008 series-quality management standards. Milwaukee, WI: ASQ Quality Press; 2008.

[7] Accreditation program in blood bank; American Association for Blood banks. http://www.aabb.org/sa/overview/Pages/program.aspx.

[8] Juran JM, Godfrey AB. Juran's quality handbook. 5th ed. New York: McGraw-Hill; 1999.

[9] World Health Organization. Geneva, 2011b. Laboratory quality management system handbook.

[10] Laboratory Methods for the Diagnosis of Meningitis Caused by Neisseria meningitidis, Streptococcus pneumoniae, and Haemophilus influenzae. 2nd ed., 2011 [Chapter 13: Quality Control/Quality Assurance].

[11] Food and Drug Administration. Guidance for industry: blood establishment computer system validation in the user's facility. Silver Spring, MD: CBER Office of Communication, Outreach, and Development; 2013. (April, 2013).

FURTHER READINGS

[12] Fung MK (Edited.), American Association for Blood bank technical manual (18th ed.), AABB, Bethesda, 2014.

[13] Clinical and Laboratory Standards Institute. Quality management system: development and management of laboratory documents; approved guideline. 6th ed. PA (CLSI): Wayne; 2013. (PO02-A6/QMS 02-A6).

Special transfusion situations

10

INTRODUCTION

Topics to be covered in this chapter include neonatal alloimmune thrombocytopenia (NAIT), hemolytic disease of the fetus and newborn (HDFN), and platelet refractoriness. Although diverse, these subjects represent a critical mass of the transfusion medicine topics which are essential to the practice of transfusion medicine. High yield points will be highlighted in the tables at the end of each section.

NEONATAL ALLOIMMUNE THROMBOCYTOPENIA

Neonatal alloimmune thrombocytopenia (NAIT), also known as fetomaternal alloimmune thrombocytopenia (FMAIT), is the most common cause of severe thrombocytopenia ($<20 \times 10^9$/L) in newborns. With an estimated incidence of 1–1.5 in 1000 live births, it is a unique alloimmune disease often affecting the first born infant. Clinical symptoms vary but because of the adherent risk for life-threatening complications such as intracranial hemorrhage (ICH), rapid diagnosis and management are a necessity [1].

CLINICAL PRESENTATION

The diagnosis of NAIT is often unexpected; almost 50% of cases occur in the first pregnancy. Typically the diagnosis is often made after birth and clinical symptoms vary. Infants may present with mild symptoms such as thrombocytopenia, bruising, petechiae or purpura but more severe bleeding symptoms may also occur. According to the literature, major bleeding occurs in 10%–20% of untreated NAIT cases. The most severe adverse sequela is intracranial hemorrhage (ICH) which is associated with 10%–12% risk of persistent neurologic impairment and 10%–15% risk of death. ICH can occur in utero or after birth. The highest risk of postnatal ICH is the first 24 to 36 h of life, as thrombocytopenia persists after delivery. Persistent neonatal thrombocytopenia is due to the persistence of maternal immunoglobulin which has a half-life of 25 days. As with HDFN, NAIT can reoccur in subsequent pregnancies if the subsequent fetus is antigen positive. Typically the degree of thrombocytopenia is often more severe in the latter cases. The only useful predictor of severity is antenatal ICH in an older affected sibling [2].

PATHOPHYSIOLOGY

As previously alluded to NAIT is an alloimmune disease, which results in the destruction of fetal or neonatal platelets by maternal alloantibodies. These alloantibodies are directed against incompatible fetal platelet antigens inherited from the father. Maternal alloantibodies are generated due to maternal fetal hemorrhage (MFH) and are IgG in nature. Transplacental passage of the IgG alloantibodies usually begins in the second trimester and continues until birth. However, fetal platelet antigen expression has been demonstrated as early as the 16th week of gestation. As such case series have shown severe thrombocytopenia (platelet count of <20,000) to be present in 46% of fetuses with NAIT who underwent fetal blood sampling (FBS) before 24 weeks gestation [1,2].

Typically maternal alloantibodies are directed against human platelet alloantigens (HPAs), mostly HPA 1a. Retrospective analyses have demonstrated the causative antibodies were directed against HPA 1a in 79% cases. However, the rate of HPA 1b homozygosity in pregnant women has been shown to be approximately 1.6%–2.5% and of those susceptible, the actual alloimmunization rate was found to be only 6%–14%. Women with HLA class II DRB3*0101 type was more likely to be alloimmunized. The remainder of NAIT cases were shown to be due to HPA 5b (9%), HPA 1b (4%), and HPA 3a (2%). When HPA 1–5 testing was negative, further investigation into HPA 9b has proven beneficial. Although rare, this alloantigen appears to be highly immunogenic in nature [3].

MANAGEMENT

The principal management goal is to prevent or arrest bleeding by rapidly restoring the platelet count. In the postnatal setting, infants are usually transfused in the setting of active bleeding and/or to maintain a platelet count of $>30 \times 10^9$/L. The platelet count should be checked within an hour after the transfusion to document response [4].

Classical management strategies focus on transfusion of antigen negative platelets whether that be derived from allogenic donors or maternal sources is contingent upon availability. In the United Kingdom, acquisition of allogenic antigen negative platelets is not as problematic and appears to be a feasible option. However, according to the US literature, various obstacles arise in the acquisition of allogenic antigen negative donor platelets resulting in the utilization of maternal platelets. Maternal platelets should be washed to remove the offending alloantibodies from the plasma, and irradiated to prevent transfusion-associated graft versus host disease. If maternal and allogenic antigen negative platelets are not available, non-antigen negative or antigen untested platelet transfusions should not be withheld because the infant is at risk for ICH. Some investigators propose coupling this practice with infusions of intravenous immunoglobulin (IVIG) at a dose up to 1 g/kg per day for 1–2 days but this practice is controversial. A recent study showed that the transfusion of random donor platelet concentrates increased the average platelet count by approximately 70,000; as compared to a 57,000 average increase when random donor platelet concentrates and IVIG were administered. As such, the authors concluded that there was no additional observed benefit of IVIG administration.

During the antenatal period, the focus of patient management is centered monitoring the fetal platelet count. Typically, this begins at 20 weeks gestation. However, in some cases it starts as early as 12 weeks gestation. Those who are monitored include mothers with a history of a previously affected infant or those with a family history of NAIT. Transfusion goals include maintaining a fetal platelet count of >50,000 with antigen negative, irradiated platelets. In addition, weekly IVIG typically at a dose of 1 g/kg per week is given. At birth, infants are delivered by C-section [1,2,4]. Major aspects of neonatal thrombocytopenia are summarized in Table 10.1.

Table 10.1 Major Aspects of Neonatal Thrombocytopenia

Clinical presentation	• Incidence: 1–1.5 in 1000 live births • 50% of cases occur in the first pregnancy • Thrombocytopenia, bruising, petechiae, purpura, ICH • Predictor of severity is antenatal intracranial hemorrhage (ICH) in the older affected sibling
Pathophysiology	• Destruction of fetal or neonatal platelets by maternal alloantibodies • Generated due to maternal fetal hemorrhage (MFH) • IgG in nature • HPA 1a (79%), HPA 5b (9%), HPA 1b (4%), and HPA 3a (2%)
Management	Antenatal • Fetal platelet count >50 × 10^9/L • Antigen negative, irradiated platelets • Weekly IVIG (1 g/kg per week) • Cesarean section birth Postnatal • Maintain a platelet count of >30 × 10^9/L • Antigen negative or maternal platelets • Maternal platelets should be washed and irradiated • If antigen negative are unavailable, nonantigen negative or antigen untested platelet transfusions should not be withheld • Intravenous immunoglobulin (IVIG) at a dose up to 1 g/kg per day for 1–2 days

HEMOLYTIC DISEASE OF THE FETUS AND NEWBORN

Hemolytic disease of the fetus and newborn (HDFN) can be due to exposure to foreign red blood cell antigens resulting in alloimmunization of the mother, either from exposure during MFH or previous transfusion. However, natural immunization may also contribute to the pathology, such as in the case of ABO isohemagglutinins. In the setting of HDFN, the immunogenic target is present on fetal red blood cells (RBCs). The target can be ABO or non-ABO antigen in origin and depending on the antigenic target, the severity of RBC destruction can vary which directly impacts clinical presentation and management.

CLINICAL PRESENTATION AND PATHOPHYSIOLOGY

The pathophysiology of HDFN centers on destruction of fetal RBCs due to passively acquired maternal antibodies. As previously mentioned, these antibodies can be naturally occurring or maybe due to exposure of foreign RBC antigens. Naturally occurring antibodies are those of ABO origin and present in the maternal serum; whereas alloantibodies are due to exposure to foreign RBC antigens and typically develop due to MFH or transfusion. Regardless of the source, maternal immunoglobulin of the IgG class transverses the placenta and binds to fetal RBCs. These antibodies coat the fetal RBCs and mark them for destruction by the splenic macrophages [5,6]. Additionally, depending target of the antibody, that is, anti-Kell or anti-GPMur, erythroid progenitor cells may also be targeted resulting in the anemia without erythroblastosis [6].

Anemia then induces the bone marrow to produce RBCs at an accelerated rate. If the rate of hemolysis is mild, the increased bone marrow production of RBC can compensate for the loss. However, if the rate of hemolysis is severe the bone marrow may fail to meet the demands resulting in extramedullary hematopoiesis; enlargement of the liver and spleen leading to portal hypertension and hepatocellular damage, which results in hypoproteinemia. In the setting of anemia and hypoproteinemia, high output cardiac failure may cause generalized edema, effusions, ascites, and ultimately cumulates in fetal hydrops.

Hemolysis of RBCs also results in hyperbilirubinemia. During the intrauterine period this is not a problem, as bilirubin passes from the fetal circulation and is cleared by the maternal liver. However, since RBC destruction continues to occur until after birth and the immature liver of the neonatal cannot efficiently conjugate bilirubin, hyperbilirubinemia develops. If left untreated, irreversible neurologic damage, namely, kernicterus, may occur. This condition is notable for deposition of bilirubin in the basal ganglia and brain stem nuclei. Clinical manifestations include athetoid cerebral palsy, hearing problems, and psychomotor handicaps [5,6].

There are numerous factors which contribute to the severity of HDFN. These include host factors, antigenic exposure, immunoglobulin class, antibody specificity, and ABO incompatibility. Studies have shown the ability to produce antibody in response to antigenic exposure varies and appears to be genetically dependent. Additionally, not all sources of MFH equivalent of the various sources, child birth poses the greatest risk for alloantibody formation; as does RBC transfusion. As a result, some investigators have postulated that transfusion of premenopausal women should be performed with Rh cDE- and K-compatible blood. Immunoglobulin sub-class also imparts an effect as IgG_1 and IgG_3 are more efficient in RBC hemolysis. However, antibody specificity appears to be more critical [7]. According to the litera-ture, there are a wide variety of RBC alloantibodies reported to cause HDFN. Popu-lation studies demonstrate that the most potent antibodies include D, c, E and K; of these studies have shown that anti-D correlates with the highest risk of fetal mortality and morbidity. Of the non-D alloantibodies, anti-K causes severe disease in 26%, anti-c in 10%, and anti-E in 2% of pregnancies. HDFN due to ABO discrepancies, while it is the most common cause of HDFN, its clinical presentation is mild. This is imparting due to the decreased amount of ABO antigens on fetal RBCs. According to the literature, adult levels of the ABO antigens are not reached until the age of 4. Additionally, since ABO antigens are present on the placental tissue, some of the antibody is absorbed by the placenta lessening the effect on the fetal RBCs [5–7]. Additionally, mothers that are type O blood group are more likely to develop HDFN during pregnancy as type O patients are more likely to have naturally occurring ABO antibodies that are of the IgG class.

MANAGEMENT

During the first prenatal visit, all pregnant women should be tested for blood type (ABO, RhD) as well as testing for antibodies that detect IgG antibodies. If the antibody detection test identifies an alloantibody, the specificity should also be determined to decipher risk stratification. While certain alloantibodies (i.e., anti-I, P1, Lea, and Leb) if present are clinically insignificant thus not associated with increased risk in pregnancy, others alloantibodies (i.e., D, E, c, K) are clinically sig-nificant and require intensive perinatal monitoring. If a clinically significant alloan-tibody is identified, the first step is to determine the fetal predicted risk of inheriting the corresponding antigen. Depending on the alloantibody identified in the maternal serum, provided paternity is assured, the father's blood can be tested either sero-logically or genotypically for the corresponding antigen. If the father is negative, no additional monitoring is necessary, but if paternal heterozygosity is identified, the fetus bears a 25% risk of antigen inheritance requiring additional monitoring. If the paternal heritage is uncertain, amniocyte genetic testing or fetal DNA obtained from the maternal serum can be used to determine whether the fetus is at risk. For preg-nancies at risk for HDFN, monitoring with serial antibody titers and radiographic studies is necessary.

Antibody titers help to further stratify fetal risk. According to the literature, estab-lished critical titer levels (16–32 commonly used) correlate with risk of significant fetal anemia. However, the Kell blood group antigen is the exception, as the Kell antigen is also present on progenitor red cells. Thus, a lower titer level of 8 has been shown to correlate with significant fetal anemia due to hypoproliferation. Titers are usually performed in parallel meaning after a titer is performed the sample is frozen, when subsequent samples are collected, a titer is repeated on the previously frozen sample and compared to the original result and the new sample. If the titer remains unchanged in the previously frozen sample, any increase in titer strength noted in the new sample is reflective of ongoing immune stimulation.

Radiographic studies such as ultrasonography and Doppler velocimetry are also used to monitor at risk pregnancies. Ultrasounds are performed weekly to look for the present of ascites while Doppler velocimetry assess the velocity of flow through the middle cerebral artery in order to estimate the degree of fetal anemia. The benefit of this procedure over the previously utilized Lyle curve is that it is noninvasive. Surveillance with this type of radiographic study is usually initiated at 16–24 weeks gestation or when a critical antibody titer is reached. After initiation, measurements are performed every 2 weeks for risk stratification. Doppler reading that are >1.5 multiples of the mean (MoM) are reflective of moderate to severe fetal anemia.

When fetal anemia becomes moderate to severe, cordocentesis is performed. Cordocentesis is also known as umbilical vein sampling and is used to directly measure the fetal hematocrit level. Hematocrit levels less than 30% necessitate intrauterine transfusion (IUT). IUT is performed by inserting a needle into the umbilical vein. RBCs are infused to achieve a predetermined hematocrit level and once this process is initiated it is repeated every 2 weeks until delivery. RBC used for IUT are typically group O (compatible with both maternal and fetal blood types), leukocyte reduced, hemoglobin S negative, cytomegalovirus (CMV) safe, irradiated, and antigen negative for the antibodies present in the maternal serum. The risk of adverse reaction during IUT is 1%–3%. Potential sequela includes infection, rupture of membranes, and suppression of fetal marrow production. Aside from IUT maternal treatment with plasma exchange and/or IVIG play a role in treatment of HDFN as adjacent therapies.

Neonatal management of HDFN focuses primary on management of hyperbilirubinemia. Prior to delivery, the maternal liver conjugates excessive fetal bilirubin generated by hemolysis. However, due to the immaturity of the neonatal liver and continual ongoing hemolysis, hyperbilirubinemia may develop in the newborn. Severe hyperbilirubinemia may lead to kernicterus, a rare type of brain damage due to deposition of bilirubinemia. Phototherapy and IVIG are usually first line therapies utilized in mild cases. Once the bilirubin levels have reached a critical level, depending on gestational age, exchange transfusions are warranted. Blood selection is similar to that of IUT and when a two volume exchange is performed 90% of RBC and 50% of bilirubin is removed. Aside from hyperbilirubinemia, infants my also suffer from persistent hyporegenerative anemia which may occur for weeks after birth. Clinical signs of anemia in infants include poor feeding, increase sleep, and other cytopenias.

Management of HDFN also focuses on prevention. In Europe, primary prevention of HDFN focuses on varying degree of Rh and Kell-matched blood to women of child bearing age. However, in the United States, the focus is on secondary prevention. In such instance RhD immune globulin (RhIg) are given either in the setting of maternal fetal hemorrhage (MFH) or at 28 weeks of gestation and at delivery to prevent alloimmunization. Studies have shown that RhIg administration may decrease alloimmunization to RhD from 1.5% to 0.1%. One vial of RhIg is given at 28 weeks gestation. Additionally, one vial of RhIg is given at delivery if the initial screening test (Rosette test) is found to be negative. If the screening test is found to be positive then determination of the amount of MFH is determined by either Kleihauer-Betke testing or flow cytometry. In that instance, RhIg dosing should be determined by the volume of fetal bleed with each vial of RhIg covering 30 mL of Rh-positive whole blood. Common aspects of hemolytic disease of the fetus and newborn are summarized in Table 10.2 [8,9].

PLATELET REFRACTORINESS

Platelet transfusion refractoriness (PTR) is defined as a less than expected increase in platelet count status post transfusion of two or more ABO-matched platelet products that are less than 72 h old. Unresponsiveness to platelet transfusion can be multifactorial and maybe due to immune-mediated or nonimmune-mediated causes. Studies have shown that nonimmune causes; namely, massive bleeding, fever, sepsis, splenomegaly, disseminated intravascular coagulation (DIC), drugs, microangiopathic hemolytic coagulation (thrombotic thrombocytopenic purpura, hemolytic uremic syndrome, thrombocytopenic syndromes of pregnancy), and poor platelet storage conditions are more frequently implicated. However, there a plethora of immune-mediated causes, such as alloimmunization to human leukocyte antigens (HLA) class I antigens, human platelet antigen alloantibodies, and ABO incompatibility. This is an important distinction as the management is very different [10–12].

Table 10.2 Major Aspects of Hemolytic Disease of the Fetus and Newborn

Clinical presentation/ pathophysiology	• Destruction of fetal RBCs due to passively acquired maternal antibodies. Can be naturally occurring or maybe due to exposure of foreign RBC antigens
	• Maternal immunoglobulin of the IgG class transverses the placenta and binds to fetal RBCs
	• These antibodies coat the fetal RBCs and mark them for destruction by the splenic macrophages
	• Anemia then induces the bone marrow to produce RBCs at an accelerated rate
	• Due to extramedullary hematopoiesis enlargement of the liver and spleen may occur leading to portal hypertension and hepatocellular damage, which may result in hypoproteinemia
	• In the setting of anemia and hypoproteinemia, high output cardiac failure ensures leading to generalized edema, effusions, ascites, and ultimately cumulates in fetal hydrops

(Continued)

Table 10.2 Major Aspects of Hemolytic Disease of the Fetus and Newborn (*cont.*)

Management	• Prenatal testing

	• Maternal blood type testing and antibody detection
	  • Serologically or genotypically type the father's RBC for the corresponding antigen
	  • Amniocyte genetic testing or fetal DNA obtain from the maternal serum can be used to determine fetal RBC have the antigen
	  • Pregnancies at risk for HDFN are monitored with serial antibody titers and radiographic studies
	  • Ultrasounds are performed weekly to look for the present of ascites
	  • Doppler velocimetry assess the velocity of flow through the middle cerebral artery to estimate the degree of fetal anemia. Doppler reading that are >1.5 multiples of the mean (MoM) are reflective of moderate to severe fetal anemia

	• Prenatal management of HDFN

	• Cordocentesis directly measures the fetal hematocrit level
	  • Hematocrit levels less than 30% necessitate IUT
	  • Transfusion repeated every 2 weeks until delivery
	  • RBC used for IUT are typically group O, leukocyte reduced, hemoglobin S negative, CMV safe, irradiated, and antigen negative for the antibodies present in the maternal serum

	• Neonatal management of HDFN

	• Focuses primary on management of hyperbilirubinemia
	    - Severe cases hyperbilirubinemia may lead to kernicterus
	    - Phototherapy and IVIG are usually first line therapies
	    - Once the bilirubin levels have reached a critical level, depend on gestational age, exchange transfusions are warranted
	  • Management of HDFN also focuses on prevention
	    - RhD immune globulin is given to Rh negative mothers in the setting of maternal fetal hemorrhage (MFH) or at 28 weeks gestation and at delivery to prevent alloimmunization
	    - RhIg dosing should be determined by the volume of fetal bleed/30 mL

CLINICAL PRESENTATION

Lack of response to platelet transfusion is the common feature of both immune and nonimmune platelet refractoriness. As mentioned above multiple external factors can affect the posttransfusion platelet count, thus platelet refractoriness should only be considered in the setting of at least two instances in which the posttransfusion count was less than expected. Typically, one dose of platelets should increase the posttransfusion platelet count by 20,000–50,000.

Immune-mediated platelet refractoriness is usually suspected in the setting of immediate (>1 h status posttransfusion) unresponsiveness to transfusion in multiparous women or those with acute leukemia requiring multi-transfusion support. It is usually due to alloimmunization to class I Human Leukocyte Antigens (HLA). However, alloimmunization to human platelet antigens (HPAs) and naturally occurring ABO antibodies have also been implicated.

Nonimmune-mediated platelet refractoriness is usually present in the setting of a consumptive process such as bleeding, fever, splenomegaly, sepsis, DIC, certain medications, autoimmune conditions or microangiopathic hemolytic syndromes. Nonimmune-mediated platelet refractoriness typically presents with an appropriate response to transfusion however, thrombocytopenia usually re-ensues within 18–24 h of transfusion. Two calculations, calculation of the correct count increment (CCI) and posttransfusion platelet recovery (PPR) are utilized not only to diagnosis platelet transfusion unresponsiveness but to differentiate the etiological cause.

The PPR equation is less frequently used. It is defined as the observed increment platelet count divided the number of platelets transfused over the patient's blood volume. The incremental platelet count is determined by measuring the platelet count immediately prior to transfusion and repeating it at least 10 min but no more than 1 h after transfusion. The number of platelets transfused can be estimated approximately using the volume and type of platelet product being transfused. Typically, one dose of apheresis platelets contains 3.0×10^{11}/L platelets while each unit of pooled platelets contains 5.5×10^{10}/L. Patient blood volumes can be approximated using Gilcher's rule of 5's for blood volumes. According to this rule, blood volumes (in mL/kg) of obese, thin, normal, and muscular men are about 60, 65, 70, and 75, respectively, whereas women with the same body habitus are estimated to have blood volumes of 55, 60, 65, and 70.

The CCI is more frequently used. This calculation uses the body surface area as a surrogate for blood volume. It is defined as the increment ($\times 10^9$/L) platelet count multiplied the body surface area expressed in meters squared divided by the number of platelets transfused ($\times 10^{11}$). The Increment count and number of platelets transfused are derived in the same fashion as discussed above. Body surface area is calculated based of height and weight. However, regardless of the equation used, both formulations account for the several confounding variables which can impact the peripheral platelet incremental count after transfusion. These include infusion dose, patient blood volume, and percent of platelets sequestered by the exchangeable splenic pool. Although the exact definition varies in the literature, immune-mediated platelet refractoriness is defined as a PPR below 30% or a CCI 7500 when the posttransfusion platelet count is measured 10–60 min after transfusion. Nonimmune platelet refractoriness is defined as a PPR below 20% or CCI 5000 when the posttransfusion platelet count is acquired 18–24 h status post transfusion. In other words immune-mediated platelet refractoriness is reflective of decreased platelet recovery, whereas nonimmune platelet refractoriness is reflective of decreased platelet survival [10–15].

PATHOPHYSIOLOGY

As mentioned above platelet refractoriness can stem from nonimmune and immune causes. Typically, nonimmune causes are more frequently implicated and consumption is the main mechanism for platelet clearance. Immune-mediated platelet refractoriness is due to alloimmunization or the presence of naturally occurring ABO antibodies. Regardless of the underlying immunologic cause, the basic mechanism is the same; when antibody is bound to antigenic target on platelets it triggers platelet destruction via splenic macrophages.

In the setting of alloimmunization, alloantibodies develop secondary to exposure to a foreign antigen either by transfusion or pregnancy. HLA class I, specifically group A and B antigens are the most common alloimmune target. However, HPA alloimmunization may also occur but this antigen is less common.

Naturally occurring ABO antibodies are generated in response to environmental antigenic stimulation. As a result, when ABO mismatched platelets are transfused this results in increased platelet destruction due to major and/or minor ABO incompatibility, particularly in individuals with high existing or reactive isoagglutinin titers (>1:256). Major incompatibility refers to preexisting isoagglutinins present in the patient's serum binding to ABO antigens present on transfused platelet. Minor ABO incompatibility refers to alloantibodies present in the plasma of the transfused platelet product binding to existing ABO antigens on platelets present in the patient's serum [13].

MANAGEMENT

As mentioned above management depends on the etiologic cause. In the setting of nonimmune platelet refractoriness the main focus of management is to address the underlying causes of thrombocytopenia, that is, alleviate bleeding, sepsis, DIC, fever. Aside from treating the underlying cause it is also recommended to transfuse the patient with the freshest ABO identical platelets. This recommendation is based on the observation from the landmark TRAP (trial to reduce alloimmunization to platelets) study where authors concluded that transfusion with fresher, ABO identical platelets resulted in higher posttransfusion CCIs [15]. Other studies have also confirmed this finding. According to another study, exclusive use of ABO-identical units resulted in 41% fewer transfusions and halved the rate of refractoriness in multi-transfused patients. Thus, this strategy appears to be beneficial in both immune and nonimmune platelet refractoriness [11].

The workup and management for immune mediate platelet refractoriness is more complex. When immune-mediated platelet refractoriness is suspected, laboratory testing for HLA antibodies should be performed. The classic assay utilized for detection of HLA antibodies is lymphocytotoxicity test (LCT). This tests, which detects IgM and IgG antibodies, reacts patient serum with a large panel of typed lymphocytes to detect cytotoxic antibodies. It is often modified by the addition of human antiglobulin. If the panel of lymphocytes is chosen carefully the percentage of positive reactions will correlate with the breadth of the patient's alloimmunization. However, due manual nature of this test as well as issues of interferences, in most clinical

Table 10.3 Various Aspects of Platelet Refractoriness

Clinical presentation	• Platelet transfusion refractoriness (PTR) less than expected increase in platelet count status post transfusion of two or more ABO matched platelet products that are less than 72 h old
	• Immune-mediated platelet refractoriness is usually suspected in the setting of immediate (>1 h status post transfusion) unresponsiveness to transfusion in multiparous women or those with acute leukemia requiring multi-transfusion support
	• Nonimmune-mediated platelet refractoriness is usually present in the setting of a consumptive process such as bleeding, fever, splenomegaly, sepsis, DIC, certain medications, autoimmune conditions or microangiopathic hemolytic syndromes. Nonimmune-mediated platelet refractoriness typically presents with an appropriate response to transfusion however, thrombocytopenia usually reensues within 18–24 h of transfusion
	$$CCI = (m^2 \times 10^9 / l \, per \, 10^{11}) = \frac{increment \, (\times 10^9 / l) \times BSA \, (m^2)}{Number \, of \, platelets \, transfused \, (\times 10^{11})}$$
	$$CCI = (m^2 \times 10^9 / l \, per \, 10^{11}) = \frac{increment \, (\times 10^9 / l) \times BSA \, (m^2)}{Number \, of \, platelets \, transfused \, (\times 10^{11})}$$
	Immune-mediated platelet refractoriness is defined as a PPR below 30% or a CCI below 7.5 when the posttransfusion platelet count is measured 10–60 min after transfusion
	• Nonimmune platelet refractoriness is defined as a PPR below 20% or CCI below 5.0 m when the posttransfusion platelet count is acquired 18–24 h status posttransfusion
Pathophysiology	• Causes of immune-mediated platelet refractoriness
	• HLA or HPA alloimmunization
	• Naturally occurring ABO antibodies
	• Nonimmune platelet destruction is due to consumption
Management	• Management is dependent on etiological cause
	• Nonimmune platelet refractoriness
	• Address the underlying cause
	• Freshest ABO identical platelet product
	• Immune platelet refractoriness
	• HLA type the patient
	• If available while awaiting HLA typing, provide crossmatch compatible platelet products
	• Aim for HLA matches that are BU (B1U) or higher
	• HLA matched platelet must be irradiated to prevent graft versus host disease
	• Platelet source (whole blood versus apheresis) does not matter provided the product is leukocyte reduced

laboratories LCT has been replaced with enzyme-linked immunosorbent assays or miniflow platforms. HPA's are detected with solid phase red cell agglutination (SPCA) assays. This methodology uses immobilized typed platelets, patient's serum and anti-IgG-coated RBCs to detect HPA antibodies. As the incidence of platelet refractoriness secondary to HPAs is low (less than 10%) and the diagnostic testing is difficult, HPA testing should be performed in the setting of serial unexplained HLA match platelet product failures.

If HLA and/or HPA alloantibodies are detected different methodologies are available for platelet selection. These included HLA-matched platelets, platelet crossmatching, and selection of antigen negative units. With HLA-based selection it is important to remember that the closet match obtainable is not always an exact match. Thus the response to platelet transfusion will be impacted by lower grade matches. Additionally, this does not address HPA antigens and the product will have to be irradiated to prevent graft versus host disease. Selection of antigen negative units is also useful in the setting of HLA platelet refractoriness. In this instance, platelet products are selected based on the specificity of the patient's antibodies. This methodology is considered to be as effective as HLA-matched and it broadens the potential number of compatible product. Platelet crossmatching is the only methodology allows for matching of both HLA and HPA antibodies. As it utilizes an SPCA platform, a technology difficult to perform and as such is not available at most institutions. Additionally, most hospital based blood banks do not have adequate supply of platelet product to achieve perfect match [11–15]. Various aspects of platelet refractoriness are summarized in Table 10.3.

KEY POINTS

- The diagnosis of NAIT is often unexpected; almost 50% of cases occur in the first pregnancy. Infants can present with mild symptoms such as thrombocytopenia, bruising, petechiae or purpura. Or severe symptoms such as ICH.
- NAIT can reoccur in subsequent pregnancies if the subsequent fetus is antigen positive. Typically the degree of thrombocytopenia is often more severe in the latter cases.
- Maternal alloantibodies directed against HPAs, mostly HPA 1a (79% cases) is responsible for destruction of fetal or neonatal platelets.
- The principal management goal is to prevent or arrest bleeding by rapidly restoring the platelet count. In the postnatal setting, infants are usually transfused in the setting of active bleeding and/or to maintain a platelet count of $>30 \times 10^9$/L.
- Classical management strategies focus on transfusion of antigen negative platelets whether that be derived from allogenic donors or maternal sources is contingent upon availability. Maternal platelets should be washed to remove the offending alloantibodies from the plasma, and irradiated to prevent transfusion-associated

graft versus host disease. If maternal and allogenic antigen negative platelets are not available, nonantigen negative or antigen untested platelet transfusions may be performed.

- The pathophysiology of HDFN centers on destruction of fetal RBCs due to passively acquired maternal antibodies. These antibodies can be naturally occurring or maybe due to exposure of foreign RBC antigens.
- Anemia induces the bone marrow to produce RBCs at an accelerated rate which may lead to extramedullary hematopoiesis; enlargement of the liver and spleen leads to portal hypertension and hepatocellular damage, which results in hypoproteinemia. High output cardiac failure may cause generalized edema, effusions, ascites, and ultimately cumulates in fetal hydrops.
- Hemolysis of RBCs after birth results in hyperbilirubinemia which can lead to irreversible neurologic damage, namely, kernicterus which is characterized by deposition of bilirubin in the basal ganglia and brain stem nuclei and clinical manifestations including athetoid cerebral palsy, hearing problems, and psychomotor handicaps.
- Antibody titers help to further stratify fetal risk. According to the literature, established critical titer levels (16–32 commonly used) correlate with risk of significant fetal anemia.
- Radiographic studies used to monitor at risk pregnancies include ultrasonography and Doppler velocimetry. Ultrasounds are performed weekly to look for the present of ascites while Doppler velocimetry assess the velocity of flow through the middle cerebral artery to estimate the degree of fetal anemia. When fetal anemia becomes moderate to severe, cordocentesis is performed.
- RBC used for IUT are typically group O (compatible with both maternal and fetal blood types), leukocyte reduced, hemoglobin S negative, CMV safe, irradiated, and antigen negative for the antibodies present in the maternal serum.
- Management of HDFN focuses on prevention. RhD immune globulin (RhIg) is given either in the setting of maternal fetal hemorrhage (MFH) or 28 weeks gestation as well as after delivery to prevent alloimmunization.
- PTR is defined as less than expected increase in platelet count status post transfusion of two or more ABO-matched platelet products that are less than 72 h old. Unresponsiveness to platelet transfusion can be multifactorial and maybe due to immune-mediated or nonimmune-mediated causes. Studies have shown that nonimmune causes are more frequently implicated.
- Immune-mediated platelet refractoriness is usually suspected in the setting of immediate (>1 h status posttransfusion) unresponsiveness to platelet transfusion. It is usually due to alloimmunization to class I HLA.
- Nonimmune-mediated platelet refractoriness typically presents with an appropriate response to transfusion however, thrombocytopenia usually reensues within 18–24 h of transfusion.

- Two calculations, calculation of the correct count increment (CCI) and posttransfusion platelet recovery (PPR) are utilized not only to diagnosis platelet transfusion unresponsiveness but to differentiate the etiological cause. The CCI is more frequently used. It is defined as the increment ($\times 10^9$/L) platelet count multiplied the body surface area expressed in meters squared divided by the number of platelets transfused ($\times 10^{11}$).
- Immune-mediated platelet refractoriness is defined as a PPR below 30% or a CCI below 7.5 when the posttransfusion platelet count is measured 10–60 min after transfusion. Nonimmune platelet refractoriness is defined as a PPR below 20% or CCI below 5.0 m when the posttransfusion platelet count is acquired 18–24 h status post transfusion.
- Immune-mediated platelet refractoriness is due to alloimmunization or the presence of naturally occurring ABO antibodies. HLA class I, specifically group A and B antigens are the most common alloimmune target.
- Nonimmune platelet refractoriness the main focus of management is to address the underlying causes of thrombocytopenia. Aside from treating the underlying cause it is recommended to supply the patient with the freshest ABO identical platelets available.
- When immune-mediated platelet refractoriness is suspected laboratory testing for HLA antibodies should be performed. If HLA alloantibodies are detected different methodologies are available for platelet selection. These included HLA-matched platelets, platelet crossmatching, and selection of antigen negative units.

REFERENCES

[1] McQuilten ZK, Wood EM, Savoia H, Cole S. A review of pathophysiology and current treatment for neonatal alloimmune thrombocytopenia (NAIT) and introducing the Australian NAIT registry. Aust N Z J Obstet Gynaecol 2011;51:191–8.
[2] Roberts I, Murray NA. Neonatal thrombocytopenia. Semin Fetal Neonatal Med 2008;13:256–64.
[3] Stachurska A, Fabijanska-Mitek J, Debska M, Muzyka K, et al. Quantitative fetomaternal hemorrhage assessment with the use of five laboratory tests. Int J Lab Hematol 2016;38:419–25.
[4] Chakravorty S, Murray N, Roberts I. Neonatal thrombocytopenia. Early Hum Dev 2005;81(1):35–41.
[5] Delaney M, Matthews DC. Hemolytic disease of the fetus and newborn: managing the mother, fetus, and newborn. Hematol Am Soc Hematol Educ Program 2015;2015:146–51.
[6] de Hass M, Thurik FF, Koelewijin JM, van der Schoot CE. Haemolytic disease of the fetus and newborn. Vox Sang 2015;109:99–113.
[7] Velkova E. Correlation between the amount of Anti-D antibodies and IgG subclasses with severity of haemolytic disease of foetus and newborn. Open Access Maced J Med Sci 2015;2:293–7.

[8] Papantoniou N, Sifakis S, Antsaklis A. Therapeutic management of fetal anemia: review of standard practice and alternative treatment options. J Perinat Med 2013;41:71–82.

[9] Bowman J. The management of hemolytic disease in the fetus and newborn. Semin Perinatol 1997;21:39–44.

[10] The Trial to Reduce Alloimmunization to Platelets Study Group. Leukocyte reduction and ultraviolet B irradiation of platelets to prevent alloimmunization and refractoriness to platelet transfusions. N Engl J Med 1997;337:1861–9.

[11] Vassallo RR. New paradigms in the management of alloimmune refractoriness to platelet transfusions. Curr Opin Hematol 2007;14(6):655–63.

[12] Bakchoul T, Bassler D, Heckmann M, Thiele T, et al. Management of infants born with severe neonatal alloimmune thrombocytopenia: the role of platelet transfusions and intravenous immunoglobulin. Transfusion 2014;54:640–5.

[13] Sacher R, Kickler T, Schiffer C, Sherman L, et al. Management of patients refractory to platelet transfusion. Arch Path Lab Med 2003;127:409–14.

[14] Schiffer CA. Diagnosis and management of refractoriness to platelet transfusion. Blood Rev 2001;15:175–80.

[15] Jackman R, Deng X, Bolgiano D, Lebedeva M, et al. Low-level HLA antibodies do not predict platelet transfusion failure in TRAP study participants. Blood 2013;121:3261–6.

Pharmacologic agents in transfusion medicine 11

INTRODUCTION

Transfusion medicine physicians are increasingly becoming involved in the management of bleeding patients. In our institution we provide a service called hemotherapy where clinical pathologists are called upon to manage actively bleeding patients in the Heart and Vascular Institute. In addition to routine blood components, various drugs are also used in the management of bleeding patients. The clinical pathologists use various agents such as protamine, desmopressin (DDAVP), Prothrombin complex concentrate (PCC), antifibrinolytic agents and recombinant factor VII. All of these agents are used with the goal of reducing bleeding. In addition Rh immunoglobulins are available from blood banks to be administered in specific situations. In-depth discussion on antiplatelets and anticoagulant agents is beyond the scope of this book.

OVERVIEW OF REVERSAL AGENTS

From time to time we may be required to reverse effects of certain drugs and this is especially so if the patient is actively bleeding or the patient is being made ready for surgery. The situation becomes urgent when emergency surgery is required. Examples of drugs which may require reversal include:

- Warfarin
- Heparin (unfractionated and LMW heparin)
- Antiplatelets (P2Y12 inhibitors and Gp IIb/IIIa inhibitors)
- Direct thrombin inhibitors (bivalirudin and dabigatran)
- Anti-Xa inhibitors (rivaroxaban and apixaban)

Warfarin is commonly used in clinical practice and is a vitamin K antagonist. Reversal of effect of warfarin includes:

- Administration of vitamin K,
- Use of fresh frozen plasma (FFP),
- PCC.

Unfractionated heparin binds to antithrombin III and the effect of heparin can be reversed by protamine. Protamine can also reverse effect of low molecular weight heparin. For reversing effect of antiplatelets use of DDAVP and/or transfusion with platelets is indicated. For correcting effects of thrombin inhibitors there is no established reversal strategy but half-life of these agents are short and discontinuation of therapy may be sufficient for reversing their effects. Hemodialysis is effective in removing dabigatran due to its low protein binding. Approximately 50% of dabigatran may be removed with 2 h of dialysis. For reversing effects of factor Xa inhibitors, discontinuation of therapy alone may be sufficient.

Recombinant factor VIIa was introduced in 1980s and is an excellent agent for prevention and treatment of severe bleeding. This agent is used primarily in treating patients with congenital or acquired hemophilia but currently it has many other uses. Recombinant factor VIIa is effective in reversing effects of synthetic polysaccharides such as fondaparinux.

REVERSING EFFECT OF WARFARIN

The coumarin type anticoagulants such as warfarin, phenprocoumon, and acenocoumarol have been used as anticoagulants for a long period of time but warfarin (coumadin) is the most commonly used drug in this category. Warfarin is a synthetic compound first developed at the Wisconsin Alumni Research Foundation in 1947 and hence the name warfarin. Currently, warfarin, a water soluble drug, is available as a racemic mixture of 50% R-warfarin and 50% S-warfarin sodium salt. The S-warfarin is four times more potent than R-warfarin. Vitamin K is essential for activation of various clotting factors (II, VII, IX, and X) and in this process vitamin K is oxidized to vitamin K epoxide. Through the action of enzyme vitamin K epoxide reductase complex 1 (VKORC1), vitamin K epoxide is converted back into vitamin K in the liver. Warfarin and its related compounds act as vitamin K antagonists by inhibiting VKORC1 and as a result the hepatic synthesis of various blood clotting factors such as prothrombin (factor II), factor VII, factor IX, factor X are impaired. In addition, warfarin also interferes with the action of other anticoagulant proteins such as protein C and protein S [1]. Warfarin is monitored by using international normalization ratio (INR).

MANAGEMENT OF SUPRATHERAPEUTIC INR

The first and most important question for management of supratherapeutic INR is whether the patient has evidence of bleeding. If the patient is bleeding then the management consists of:

- Stop warfarin
- Vitamin K, slow IV infusion
- Fresh frozen plasma (FFP)
- If there is concern for volume overload with FFP then PCC can be considered

For most warfarin-treated patients with INR over 4 but not bleeding, oral vitamin K in dosage between 1 and 2.5 mg should bring INR down within 24 h. Intravenous vitamin K can lower INR more rapidly, but at 24 h, intravenous vitamin K and oral vitamin K show similar results. However, for patients with very high INR (>9) but not bleeding higher dosage of vitamin K (2.5–5 mg, even up to 10 mg) is recommended.

Warfarin-related bleeding is a serious medical complication as up to 10% patients may die within 30 days due to excessive bleeding. The most lethal hemorrhage is intracranial hemorrhage (50% of such patients die) but major gastrointestinal bleeding may also cause death. Fresh frozen plasma can also reverse effect of warfarin but each 250 mL of FFP produces only small augmentation in the activity of individual clotting factors. Therefore, in a patient with profound bleeding more than 1500 mL of FFP may be needed. In contrast PCC contains much higher amount of vitamin K dependent clotting factors per unit volume. Recombinant factor VII has been used off-label in patients with serious warfarin associated bleeding [2].

Warfarin therapy should be stopped 4–7 days prior to surgery. The INR prior to surgery should be <1.5. If surgery is required sooner, vitamin K, FFP or PCC may be used to reverse effect of warfarin. Dosages of these reversing agents are listed in Table 11.1.

FRESH FROZEN PLASMA

Fresh frozen plasma is used for reversing effect of warfarin because it contains substantial levels of vitamin K dependent clotting factors but requires a relatively large amount of FFP to correct INR. Presence of hemorrhage is a solid indication of using FFP. However, transfusion-related acute lung injury (TRALI), transfusion-related circulatory overload, allergic reactions from mild to anaphylactoid, acute hemolysis and transfusion-related infections are risks of FFP transfusion. Therefore, FFP is best used in high-risk patients in appropriate dosage after considering potential transfusion reactions [3].

PROTHROMBIN COMPLEX CONCENTRATE

There are various drawbacks of using FFP in reversing effect of warfarin. One factor is time delay for ABO blood typing and thawing of frozen plasma. Other factors include large volume of FFP which may be required to correct INR, longer time needed for infusion and risk of transfusion reaction and requirement. However, PCC products contain concentrated vitamin K dependent coagulation factors thus requiring small infusion volume compared to FFP. Moreover, such products can be used promptly because ABO blood typing is not required.

PCC products, derived from human plasma contain factors II, IX, and X, with variable amounts of factor VII along with anticoagulant proteins C, S, and antithrombin. In the United States, three distinct classifications of PCC products are commercially available:

- Three factors PCC products containing mostly factors II, IX, and X but may contain some factor VII (products available in the United States: Profilnine SD, Bebulin-VH)

Table 11.1 Agents Used for Reversal of Warfarin Effect

Reversing Agent	Administration	Adult Dosage	Comments
Vitamin K	Oral	1.0–2.5 mg vitamin K if INR >4.0 but patient is not bleeding. To correct higher INR and or bleeding, dosage between 2.5 and 5.0 mg may be used	Oral vitamin K is effective in lowering INR but it may take up to 24 h for its full effect which may last for days. Orally administered may also be used more than 24 h prior to surgery to correct INR if needed
Vitamin K	Infusion	2.5–5.0 mg vitamin K For serious bleeding of major warfarin overdose, 10 mg may be administered intravenously	Intravenously administered vitamin K can lower INR more quickly than orally administered vitamin K and such effects may last for days. Vitamin K may also be administered intravenously to correct INR 6–24 h prior to surgery
Fresh frozen plasma (FFP)	Infusion	Common practice is to infuse two units (400–500 mL) FFP Other recommendation:10–20 mL/kg	Effect is within 1 h and duration of effect is up to 12 h. For significant bleeding, large volume may be needed. FFP can also be administered <6 h prior to surgery to correct INR
K-centra	Infusion	INR: 2–4; 25 units/kg (maximum: 2500 units) INR: 4–6: 35 units/kg (maximum: 3500 units) INR: >6; 50 units/kg (maximum: 5000 units)	Effect is immediate (within 5–15 min) which may last 12–24 h. K-centra is superior to FFP because such preparation contains much higher levels of vitamin K-dependent clotting factors

- Four factors PCC product containing factors II, VII, IX, and X (product available in the United States: K-centra, approved by the FDA in 2013)
- Activated PCC (aPCC) product contains four coagulation factors (in inactive and active form).

In general four factors PCC such as K-centra (only FDA approved product available in the United States) is superior to three factors PCC. Currently, one aPCC product is available in the United States, which is a freeze dried sterile human plasma fraction with factor VIII inhibitor bypassing activity). The product is available either as FEIBA NF (antiinhibitor coagulant complex nanofiltered) and as FEIBA

VH (antiinhibitor coagulant complex vapor-heated) both manufactured by Baxter. Although aPCC is approved by the FDA to control spontaneous bleeding episodes and to prevent bleeding during surgery in hemophilia patients A and B with inhibitors to factors VIII and IX, respectively, recent data have suggested that this product may be used off-label as an anticoagulant reversal agent. However, thrombotic complication such as venous thromboembolism, disseminated intravascular coagulation (DIC), myocardial infarction, and pulmonary embolism (PE) have been reported when aPCC products have been administered for anticoagulant reversal. However, when aPCC product is used for its FDA approved indication (hemophilia), only 4–9 thrombotic events per 100,000 infusions have been reported [4].

PCC products are prepared by solvent detergent treatment of human plasma which inactivates enveloped viruses (HIV, hepatitis B, hepatitis C, human T-lymphotropic virus, etc.). As a result there is no risk of viral transmission due to PCC infusion. Moreover, additional purification step by manufacturers also significantly eliminates risk of transmission on non-enveloped viruses such as hepatitis A virus and Parvovirus B19. As mentioned earlier, four factors PCC product such as K-centra is superior to three factors PCC products in reversing effect of warfarin. K-centra contains clotting factors II, VII, IX, and X as well as protein C and S. K-centra can be reconstituted quickly and can be infused in small volume to reduce INR rapidly.

Although K-centra is approved by FDA for treatment of warfarin-related coagulopathy, it has also been used to treat coagulopathy due to use of non-vitamin K antagonist oral anticoagulant such as apixaban, dabigatran, and rivaroxaban. In the trial submitted to the FDA to obtain approval of K-centra, data showed that K-centra reduced INR to ≤ 1.3, 30 min after the end of infusion in 62.2% patients on warfarin therapy compared to 9.6% patients who received FFP. In one study based on 187 patients with elevated INR due to warfarin therapy (128 patients) or non-warfarin related therapy (61 patients), the authors observed that administration of PCC products corrected INR to 1.3 or less in 53.9% patients receiving warfarin and 27.7% patients in the non-warfarin group. Moreover, authors also reported that K-centra was more effective than three factors PCC because infusion of three factors PCC reduced mean INR from 4.64 to 1.85 in warfarin-treated patients but infusion of K-centra reduced mean INR from 4.54 to 1.30 in warfarin-treated patients indicating that K-centra reduced the mean INR by a greater absolute value than three factors PCC. In addition K-centra may also be effective in lowering INR due to administration of apixaban, dabigatran, and rivaroxaban but these uses are off-label use of K-centra and dosage must be at the discretion of clinicians [5].

Dosage of K-centra is based on units (international units) of factor IX per kg body weight. The potency of factor IX is stated on the vial of K-centra and it usually varies between 20 and 31 factor IX units/mL.

- If INR is between 2 and 4, dosage should be 25 units/kg (for a 70 kg male 1750 units)
- If INR is between 4 and 6, dosage should be 35 units/kg (for a 70 kg male 2450 units) and

- If INR is over 6, dosage should be 50 units/kg (for a 70 kg male 3500 units).

However, for dosage calculation, maximum weight should be considered as 100 kg even in a patient weighing over 100 kg. Therefore, if INR is between 2 and 5, maximum dosage should be 2500 units, if INR is between 4 and 6, maximum dosage should be 3500 units and for INR exceeding 6, maximum dosage should be 5000 units (Table 11.1).

K-centra should be administered only once and vitamin K administration may be necessary 10–12 h later when effect of K-centra is weaning off. Repeated doses of K-centra do not improve efficacy and increases risk of thromboembolic complications. K-centra should be infused through a separate line. The maximum rate of IV infusion should be 210 units/min.

PT/PTT should be checked 2 h after K-centra administration to check for therapeutic effect.

K-centra contains heparin and is contraindicated in patients with heparin-induced thrombocytopenia (HIT). Other contraindications of using K-centra include:

- Suspected disseminated intravascular coagulation (DIC)
- Acute myocardial infarction, acute septicemia, acute crush injury, acute peripheral arterial occlusion, acute thrombotic stroke, acute deep vein thrombosis (DVT) or pulmonary embolism (PE) (within 3 months), or high risk thrombophilia
- Lupus anticoagulant/anticardiolipin antibodies
- Protein C, protein S, or antithrombin deficiency
- Homozygous factor V Leiden
- Double heterozygous (factor V Leiden/ factor II G20210A prothrombin mutation)
- Pregnancy
- In the past 30 days: history of transient ischemic stroke, angina pectoris, or limb claudication

REVERSING EFFECTS OF HEPARIN AND LOW-MOLECULAR WEIGHT HEPARIN

Heparin is a natural occurring mixture of various polysaccharides with different molecular weights that is present in human and animal tissues most commonly liver and lungs. Heparin was originally isolated from canine liver and the name was derived from the Greek word "hepar" meaning liver. Commercially available heparin is isolated from bovine or porcine source and crude heparin requires purification before use. Unfractionated heparin is a heterogenous mixture of polysaccharides with molecular weight varying from 5000 to 30,000 Daltons and number of saccharide units in heparin molecules varies from 5 to 35. Heparin itself has no anticoagulant effect but when heparin molecule containing 18 of more polysaccharide binds with antithrombin III (ATIII), then this complex becomes 1000 fold more potent than ATIII itself in preventing coagulation by blocking thrombin and its enzymatic conversion of fibrinogen into fibrin. Heparin also inhibit clotting by interfering with other procoagulant effect of thrombin such as thrombin induced activation of clotting factors V and VIII that accelerates clot formation. Heparin-antithrombin III complex also enhances neutralizing effects of ATIII on activated coagulation factors IX, X, XII, and XIII as well as kallikrein which contribute to clot formation. Heparin also binds to heparin cofactor II, a glycoprotein that inactivates thrombin independent of ATIII. Heparin must be administered intravenously (IV) or by subcutaneous injection and is very effective in stopping clot formation in arteries of patients with cardiovascular diseases, during cardiac surgery, preventing clot formation in patients undergoing hemodialysis or hemofiltration, treating DVT, during orthopedic surgery, neurosurgery, and treating trauma patients [6].

Low-molecular weight heparins are prepared from unfractionated heparin by either chemical depolymerization or by heparinase digestion (tinzaparin). The molecular weights vary from 4400 to 6500. The low-molecular weight heparins are parenterally administered drugs and include ardeparin, (average molecular weight 5500–65000), bemiparin (average molecular weight 3600), certoparin (average molecular weight 5400), dalteparin (average molecular weight 6000), enoxaparin (average molecular weight 4500), nadroparin (average molecular weight 4300), parnaparin (average molecular weight 5000), reviparin (average molecular weight 4400) and tinzaparin (average molecular weight 6500). Compared to unfractionated heparin, low-molecular weight heparins show more predictable dose-response curve. Low-molecular weight heparins can be administered as fixed dosage based on body weight and they reach peak level 2–4 h after subcutaneous administration. Average half-life is 3–4 and approximately 80% drugs are eliminated via renal route. Unfractionated heparin therapy should be monitored by measuring a PTT while therapy with low-molecular weight heparin should be monitored by using antifactor Xa assay. Only reversal agent for heparin approved by the FDA is protamine.

PROTAMINE

Protamines are polycationic, highly basic polypeptides which are required for spermatogenesis in vertebrates including human. Commercially available protamine is usually prepared from gonad of male salmon. Despite the low-therapeutic index and side effects protamine is the only approved antidote of heparin overdose [7]. Protamine is also combined with insulin to increase its duration of action, for example, NPH insulin (Neutral Protamine Hagedorn also known as Humulin N, Novolin N among others) and insulin lispro protamine/insulin lispro.

Pharmacological effect of both unfractionated heparin and low-molecular weight heparin can be reversed by using protamine sulfate. In general, 1 mg protamine sulfate given intravenously will neutralize 100 units of heparin given in previous 4 h. Protamine neutralizes all antithrombin effects of low-molecular weight heparins but incompletely reverses factor Xa inhibition. In general, 1 mg of protamine given intravenously will neutralize 1 mg of enoxaparin or 100 units of dalteparin as well as 100 units of tinzaparin given within 8 h. A second dosage of 0.5 mg protamine may be administered if bleeding continues.

Heparin rebound phenomenon is defined as reappearance of hypocoagulability after adequate neutralization of heparin by protamine sulfate. Heparin rebound may occur in the early postoperative period of cardiac surgery and such phenomenon may not be detected by monitoring activated coagulation time (ACT) because ACT cannot detect residual heparin activity. So, question remains how we can effectively demonstrate presence of circulating heparin in a bleeding patient:

- TEG (thromboelastography) analysis with or without heparinase may allow diagnosis of heparin rebound [8]. In patients with circulating heparin the R time with TEG without heparinase is prolonged. In contrast, if the TEG is repeated with heparinase the R time shortens by at least 50%.
- Presence of circulating heparin may also be detected by reviewing results of a DIC panel. A DIC panel consists of PT, PTT, TT (thrombin time), fibrinogen, and D-dimer. Individuals who have circulating heparin will typically have prolonged PTT and TT with normal fibrinogen. Two common causes of prolonged TT are heparin effect and hypofibrinogenemia. If the fibrinogen level is normal then the most likely cause of prolonged TT is heparin effect.

Precise mechanism of heparin rebound is not understood. This phenomenon may be due to delayed release of heparin that is previously sequestered in tissues (especially in adipose tissue in obese patients) into circulation. In this case additional protamine is given to neutralize heparin. Heparin may also be released from heparin–protamine complex causing heparin rebound phenomenon.

RECOMBINANT FACTOR VIIa

Recombinant factor VIIa (trade name NovoSeven, also known as Eptacog Alfa) is a vitamin K dependent glycoprotein consisting of 406 amino acid residues (molecular weight 50 K Da) which is structurally similar to human plasma derived coagulation factor VIIa. Recombinant factor VIIa is manufactured using DNA biotechnology where the gene for human factor VII is cloned and expressed in baby hamster kidney cells cultured in a media containing calf serum as a nutrient. Recombinant FVII is secreted into the culture media which is purified by a chromatographic process. No human serum or other proteins are used in the production or formulation of NovoSeven. When reconstituted with sterile water for infusion, each vial contains 600 μg/mL of recombinant factor VIIa.

Recombinant factor VIIa is approved by the FDA for the treatment of bleeding episode or for prevention of bleeding during invasive procedure or surgery in patients with congenital or acquired hemophilia with antibodies toward factor VIII (hemophilia A) and factor IX (hemophilia B). At normal physiological concentrations, the procoagulant response cannot proceed to complete hemostasis without factor VIII and IX. In patients with hemophilia factor VIII or factor IX are deficient. At high concentration, recombinant factor VIIa acts by bypassing factor VIII/IX coagulation pathway either by tissue factor independent (recombinant factor VIIa can bind to activated platelet in the absence of tissue factor thus activating factor X to Xa sufficiently to mediate a burst of thrombin generation for homeostasis) or tissue factor dependent (recombinant factor VIIa in the presence of limited amount of tissue factor overcomes the inhibitory effect of zymogen factor VII on thrombin generation) pathway [9]. The recommended dosage is 90 µg/kg administered as bolus at intervals of 2–3 h for treatment of bleeding episodes in hemophiliac patients (Table 11.2).

Interestingly, according to one report, 97% use of recombinant factor VIIa in hospital settings are off-label use while management of bleeding episodes in hemophiliac patients accounts for only 3% use. The off-label use of recombinant factor VIIa include treating patients with factor VII deficiency (inherited or acquired), von Willebrand disease and Glanzmann's thrombasthenia. The three most common off-label uses of recombinant factor VIIa include management of bleeding during cardiac surgery, trauma patients, and patients with intracranial hemorrhage which together account for 69% of use of this product in hospitals in 2008. Remaining use is dispersed among various other medical and surgical indications [10]. Recombinant factor VIIa is also used off-label for reversal of effect of warfarin or newer anticoagulant drug fondaparinux. Off-label uses of recombinant factor VII are summarized in Table 11.3.

DESMOPRESSIN

Desmopressin (1-deamino-8 D-arginine vasopressin, DDAVP) is a synthetic analog of antidiuretic hormone (L-arginine vasopressin) which was first used in 1977 to treat patients with hemophilia A and von Willebrand disease, the most frequent of congenital bleeding disorders. However, desmopressin is most effective in patients with mild hemophilia A (factor VIII activity over 5%) and patients with type 1 von Willebrand disease. This drug is now listed as an "essential drug" by the World Health Organization (WHO). FDA approved desmopressin in February 1978. This drug is capable of increasing plasma levels of factor VIII and von Willebrand factor (vWF) in blood probably by releasing these factors from storage sites, for example, releasing vWF from vascular endothelium.

Desmopressin is available in intravenous, subcutaneous, intranasal, and oral formulation. Desmopressin was first used during dental extraction and then during major surgical procedures in patients with mild hemophilia or von Willebrand disease. This drug is also used for treating bleeding in patients with platelet dysfunction including uremic thrombocytopathia, thrombocytopathia due to

Table 11.2 FDA Approved Use of Other Agents for Management of Bleeding or Use as Reversal Agent

Agent	FDA Approved Use	Administration	Adult Dosage	Comment
Protamine	Reversing effect of heparin and low molecular weight heparin	Infusion	1 mg given neutralizes 100 units of heparin	Effect observed within 5 min and duration of effect is dose-dependent. In cardiac surgery heparin rebound may occur in 2–4 h
Recombinant factor VIIa	Treating bleeding in hemophilia A and B patients	Infusion	90 μg/kg for treating bleeding in hemophilia patients but lower dosage (15–30 μg/kg) for 4–6 h until hemostasis is achieved in patients with factor VII deficiency (off-label use)	Effect within 5–10 min and duration of effect is 4–6 h
Desmopressin	Treating patients with mild hemophilia A and von Willebrand disease, other bleeding disorder and diabetes insipidus. Nasal spray approved in 2017 for treating frequent urination at night	Infusion, oral, subcutaneous, and intranasal	For IV or subcutaneous dose is 0.3 μg/kg but some guideline suggests maximum dose of 20 μg to avoid side effects. Intranasal dose 150 μg (one puff) for patients weighing less than 50 kg and 300 μg (two puff) for patients weighing over 50 kg	After IV infusion, peak activity is observed in 30–60 min but after intranasal or subcutaneous administration 90–120 min

Table 11.2 FDA Approved Use of Other Agents for Management of Bleeding or Use as Reversal Agent (*cont.*)

Agent	FDA Approved Use	Administration	Adult Dosage	Comment
RhoGAM	Preventing alloimmunization during pregnancy. Prevention of D-sensitization after mismatched transfusion	Intramuscular	For preventing alloimmunization in 300 µg at 26–28 week of gestation and 300 µg within 72 h of childbirth pregnancy. For mismatched platelet transfusion 125 or 300 µg	Very effective with a long half-life and as a result a single dose lasts 2–4 weeks
Aminocaproic acid	Treating fibrinolytic bleeding. Orphan use: treating recurrent hemorrhage in patients with traumatic hyphema	Infusion and oral	For infusion 4–5 gm within first h followed by continuous infusion of 1 g per h maximum 8 h. For oral administration 5 g in the first hour and then 1 g per h till bleeding is controlled up to 8 h	Contraindicated in patients with DIC. Half-life is approximately 2 h and 65% of the drug is excreted unchanged in urine. Adipic acid is the minor metabolite
Tranexamic acid	Short term use (2–8 days) as an injection to reduce or prevent bleeding during tooth extraction in patients with hemophilia. Oral formulation approved for treating menorrhagia	Infusion, oral, and topical	Typical dosing 10 mg/kg intravenously given 3–4 times daily for 2–8 days. Oral recommended dosage is two 650 mg tablets three times a day for a maximum total daily dosage of 3900 mg for a maximum of 5 days	Mostly excreted in urine unchanged. Half-life is approximately 2 h. Effect may last 8–12 h

Table 11.3 Off-label Use of Recombinant Factor VIIa, Aminocaproic Acid, and Tranexamic Acid

Drug	Common Off-label Use
Recombinant factor VIIa	Adult and pediatric cardiac surgery, and other vascular procedures, liver transplant, trauma patients, intracranial hemorrhage, pulmonary hemorrhage, aortic aneurysm, obstetrical hemorrhage, neurosurgery, prostatectomy as well as treating primary and secondary clotting disorders
Aminocaproic acid	Reducing postoperative blood loss in cardiac, spine, orthopedic and other surgeries, preventing recurrence of subarachnoid hemorrhage, reduce bleeding in patients with severe thrombocytopenia, prevent of oral bleeding in patients with congenital and acquired coagulation disorders, and reduction of the risk of catastrophic hemorrhage in patients with acute promyelocytic leukemia. Aminocaproic acid is also used as a prophylaxis in patients with hereditary angioedema
Tranexamic acid	Reducing/preventing blood loss during cardiac surgery, spinal, or cranial surgery, total hip replacement, knee replacement and related orthopedic surgeries. It is used in trauma patients to reduce blood loss as well as in preventing blood loss during cesarean section and vaginal delivery. This drug is also used in preventing blood loss during dental procedures and oral surgeries as well as a prophylaxis in patients with hereditary angioedema

antiplatelet medications, and vascular abnormalities such as Ehlers–Danlos syndrome. Desmopressin may be effective in treating acquired bleeding disorders including drug induced bleeding disorders. Menorrhagia, defined as passage of more than 80 mL of blood per month, is the most common complications that affect women with congenital bleeding disorder. Desmopressin is an alternative approach to treat patients who cannot take hormonal therapy or are actively attempting to be pregnant [11].

Standard dosages of desmopressin for treating hemophilia and von Willebrand disease are listed in Table 11.2. After intravenous administration, peak activity is observed 30–60 min (for subcutaneous or intranasal administration, peak in 90–120 min) where levels of factor VIII and vWF are increased 3–5 times [12]. The half-life of desmopressin is 1.5–2.5 h but may be longer in renally compromised patients. The half-life of homeostasis effect is 6–8 h. On March 3, 2017 FDA approved desmopressin nasal spray (Noctiva) for treatment of nocturia due to polyuria in adults who awaken at least two times per night to void. This common condition affects over 89 million Americans. However, active pharmaceutical ingredient in Noctiva is a slight modification of desmopressin acetate (serenity) which was originally approved by the FDA in 1978. This modification was made to increase absorption after intranasal administration [13]. Desmopressin is an analog of antidiuretic hormone and is also used for treating diabetes insipidus.

RHOGAM

Rhesus (Rh_o[D]) immune globulin is a hyperimmune plasma derivative directed at red blood cells positive for Rh antigen (D antigen). Rh alloimmunization is the most common cause of hemolytic disease of the fetus and newborn where the fetus inherits the D antigen from the father while mother is D-antigen negative. Fetal cells crossing maternal circulation stimulates maternal immune system to produce Rh (D) antibodies capable of crossing the placenta into the fetal circulation causing fetal hemolysis which may occur during pregnancy or at childbirth. The incidence of Rh incompatible pregnancies is estimated to be 9%–17%. However, alloimmunization can be prevented by using Rh_o(D) immune globulin (RhoGAM and related products) during 26–28 week of gestation when most pregnant women develop anti-D antibody in Rh incompatible pregnancy. RhoGAM administration leads to rapid binding and clearance of D-positive red blood cells from maternal circulation. RhoGAM should be administered only intramuscularly at the dosage of 300 μg at 26–28 week of gestation and 300 μg within 72 h of childbirth (Table 11.2). RhoGAM has a long half-life of average 18 days. RhoGAM is available in ultrafiltered plus formulation containing 300 μg in one vial but a smaller dosage of 50 μg (MICRhoGAM) is also available. This product also may be used for obstetrical conditions such as antepartum fetal–maternal hemorrhage, actual or threatened pregnancy loss and ectopic pregnancy.

In addition to RhoGAM, three other Rh_o(D) immune globulin products are commercially available in the United States, also for preventing alloimmunization during pregnancy and as a result D-antigen sensitization rate is virtually nonexistence (0–2.2) as reported by multiple studies of at-risk women. Although RhoGAM is not approved for treating idiopathic thrombocytopenia purpura (ITP), two other immune globulin products WinRho, and Rhophylac are approved for treating ITP. Both products can be administered by intravenous or intramuscular route. However, another product HyperRHO can be only administered intramuscularly and is not approved for treating ITP [14].

As expected, another application of Rh_o(D) immune globulin products is prevention of D-sensitization after mismatched transfusion of blood components. Studies and case reports over the past 40 years have demonstrated that alloimmunization of mismatched RBC transfusion can be successfully prevented using Rh_o(D) immune globulin products when given within 72 h. One dose of 300 μg is expected to suppress immunization potential of 15 mL D antigen positive RBC or 30 mL of D+ whole blood. Given the small amount of D+ positive RBC that may be present in platelets, 125 or 300 μg should be sufficient to prevent alloimmunization during mismatched platelet transfusion [15].

ANTIFIBRINOLYTIC AGENTS

There are three available antifibrinolytic agents-aprotinin, aminocaproic acid and tranexamic acid. Aprotinin is a serine protease inhibitor while both aminocaproic acid and tranexamic acid are lysine analogs which suppress fibrinolysis by inhibiting plasminogen and the binding of plasmin to fibrin thus increasing clot stability. However, aprotinin is very toxic and as a result both aminocaproic acid and tranexamic acid are widely used in clinical practice.

Antifibrinolytic agents are used in situations of primary fibrinolysis but not in secondary fibrinolysis such as DIC. Bleeding due to primary fibrinolysis may be associated with cardiopulmonary bypass, hematological disorders such as a megakaryocytic thrombocytopenia, liver cirrhosis, and various neoplastic diseases.

AMINOCAPROIC ACID

Aminocaproic acid (epsilon amino caproic acid, 6-aminohexanoic acid; trade name: AMICAR) is FDA approved for treating bleeding due to elevated fibrinolytic activity (fibrinolytic bleeding) and also carries an Orphan drug designation from the FDA (granting special status to a drug to treat a rare disease or condition upon request of a sponsor) for prevention of recurrent hemorrhage in patients with traumatic hyphema. Off-label uses of aminocaproic acid include reducing postoperative blood loss, preventing recurrence of subarachnoid hemorrhage and preventing attacks of hereditary angioedema. This drug is water soluble and is available as tablet, oral solution and intravenous administration. Aminocaproic acid is contraindicated in patients with evidence of an active intravascular clotting process such as DIC.

For the treatment of acute bleeding syndromes due to elevated fibrinolytic activity, initial dosage should be 5 gm of aminocaproic acid (five 1000 mg tablets or equivalent oral solution) within the first hour and then 1 gm of aminocaproic acid per hour till bleeding is controlled. The treatment can be continued up to 8 h. For infusion, AMICAR 20 mL vial contains 5 gm of aminocaproic acid can be used where it has been suggested that 16–20 mL (4–5 g) of AMICAR injection in 250 mL of diluent may be administered by infusion during the first hour of treatment, followed by a continuing infusion at the rate of 4 mL (1 g) per hour in 50 mL of diluent. Such infusion can be continued till bleeding is controlled stopped up to 8 h.

Renal excretion is the primary route of elimination of aminocaproic acid regardless of route of administration. Approximately 65% of dosage is excreted in urine as the parent drug and approximately 11% as adipic acid (metabolite). The elimination half-life is approximately 2 h. AMICAR injection contains benzyl alcohol as a preservative and should not be used in newborns because benzyl alcohol has been associated with fatal "gasping syndrome" especially in premature neonates.

TRANEXAMIC ACID

Tranexamic acid was approved by the FDA in 1986 for short term use (2–8 days) as an injection to reduce or prevent bleeding during tooth extraction in patients with hemophilia. In November 2009, FDA approved oral formulation of tranexamic acid for treating menorrhagia. All other uses of tranexamic acid including preventing or minimizing postoperative blood loss, reducing blood loss in trauma patients and use as a prophylaxis in patients with hereditary angioedema are off-label use. The CRASH-2 trial (clinical randomization of antifibrinolytics in significant hemorrhage) data has clearly shown that intravenous tranexamic acid is effective in reducing mortality due to bleeding in trauma patients, but tranexamic acid must be administered within the first 3 h. Tranexamic acid is also effective in reducing bleeding during various surgery including cardiac surgeries where need for blood transfusion was significantly reduced and as a result better patient outcome is observed. Studies have also shown that tranexamic acid can reduce blood loss during caesarean section and vaginal delivery. Tranexamic acid is also effective in management of postpartum hemorrhage following vaginal delivery. Tranexamic acid may also be useful in reducing gastrointestinal bleeding and mortality. Tranexamic acid may also be applied topically for reducing bleeding from surgical wounds [16]. Off-label uses of tranexamic acid are listed in Table 11.3.

Dosing of tranexamic acid is typically 10 mg/kg intravenously given three to two times daily for 2–8 days. It is infused at a maximum rate of 1 mL per minute because more rapid infusion may cause hypotension. Tranexamic acid is also available in tablet form (trade name: Lysteda) and recommended dosage is two 650 mg tablets three times a day for a maximum total daily dosage of 3900 mg for a maximum of 5 days for treating menorrhagia. Dosing should be adjusted for renal insufficiency because the drug is mostly secreted in urine unchanged. The half-life is approximately 2 h [17].

KEY POINTS

- Warfarin and its related compounds act as vitamin K antagonists by inhibiting VKORC 1 (vitamin K epoxide reductase complex 1) and as a result the hepatic synthesis of various blood clotting factors such as prothrombin (factor II), factor VII, factor IX, factor X are impaired. In addition, warfarin also interferes with the action of other anticoagulant proteins such as protein C and protein S.
- Patients experiencing clinically significant bleeding and elevated INR due to warfarin therapy can be managed by discontinuation of warfarin therapy, administration of vitamin K and FFP or PCC.
- For most warfarin treated patients with INR over 4 but not bleeding, oral vitamin K in dosage between 1 and 2.5 mg should bring INR down within 24 h. Intravenous vitamin K can lower INR more rapidly, but at 24 h, intravenous vitamin K and oral vitamin K show similar results. However, for patients with very high INR (>9) but not bleeding higher dosage of vitamin K (2.5–5 mg) is recommended.
- Fresh frozen plasma (FFP) is used for reversing effect of warfarin because it contains substantial levels of vitamin K dependent clotting factors but requires a relatively large amount of FFP to correct INR.
- Four factors PCC products such as K-centra (only approved product in the United States) is superior to three factors PCC products for reversing effect of warfarin. The recommended dosage is for INR: 2–4; 25 units/kg (maximum: 2500 units), for INR: 4–6; 35 units/kg (maximum: 3500 units) and for INR: >6:50 units/kg (maximum: 5000 units).

- Protamine is the only FDA approved product for reversal of effect of heparin and low-molecular weight heparin. Dosage: 1 mg protamine neutralizes 100 units of heparin.
- Recombinant factor VIIa is approved by the FDA for the treatment of bleeding episode or for prevention of bleeding during invasive procedure or surgery in patients with congenital or acquired hemophilia with inhibiting antibodies toward factor VIII (hemophilia A) and factor IX (hemophilia B). However, approximately 97% of use of recombinant factor VIIa is off-label use.
- The recommended dosage of recombinant factor VIIa is 90 µg/kg for treating bleeding in hemophilia patients but lower dosage (15–30 µg/kg) for 4–6 h until hemostasis is achieved in patients with factor VII deficiency (off-label use).
- Desmopressin (DDAVP) is a synthetic analog of antidiuretic hormone (L-arginine vasopressin). Desmopressin is approved by the FDA for treating patients with mild hemophilia A and von Willebrand disease, and diabetes insipidus. On March 3, 2017 FDA approved desmopressin nasal spray (Noctiva) for treatment of nocturia due to polyuria in adults who awaken at least two times per night to void.
- Desmopressin dosage: for IV or subcutaneous dose is 0.3 µg/kg. The intranasal dose is 150 µg (one puff) for patients weighing less than 50 kg and 300 µg (two puff) for patients weighing over 50 kg.
- RhoGAM is FDA approved for preventing alloimmunization during pregnancy and also prevention of D-sensitization after mismatched transfusion. For preventing alloimmunization dosage is in 300 µg at 26–28 week of gestation and 300 µg within 72 h of childbirth pregnancy. For mismatched platelet transfusion dosage is 125 or 300 µg.
- Aminocaproic acid is approved by the FDA for treating fibrinolytic bleeding. Orphan use: treating recurrent hemorrhage in patients with traumatic hyphema. However, there are many off-label uses of aminocaproic acid including preventing or reducing blood loss during cardiac and other surgeries. This drug is contraindicated in patients with DIC.
- Tranexamic acid is approved by the FDA for short term use (2–8 days) as an injection to reduce or prevent bleeding during tooth extraction in patients with hemophilia. Oral formulation approved for treating menorrhagia. However, there are many off-label uses including preventing or reducing blood loss during cardiac and other surgeries.

REFERENCES

[1] Haustein KO. Pharmacokinetic and pharmacodynamic properties of oral anticoagulants, especially phenprocoumon. Semin Thromb Hemost 1999;25:5–11.
[2] Garcia DA, Crowther MA. Reversal of warfarin: case based practice recommendations. Circulation 2012;125:2944–7.
[3] Hartman SK, Teruya J. Practice guidelines for reversal of new and old anticoagulants. Dis Mon 2012;58:448–61.

[4] Awad NI, Cocchio C. Activated prothrombin complex concentrates for the reversal of anticoagulant-associated coagulopathy. Pharm Ther 2013;38:696–701.

[5] Mohan S, Howland MA, Lugassy D, Jacobson J, et al. The use of 3 and 4-factor prothrombin complex concentrate in patients with elevated INR. J Pharm Pract 2017; January 1, [e-pub ahead of print].

[6] Hassell K. Anticoagulation in heparin induced thrombocytopenia: an ongoing challenge. Hosp Physician 2000;36:56–61.

[7] Sokolowska E, Kalaska B, Miklosz J, Mogielnicki A. The toxicology of heparin reversal with protamine: past, present and future. Expert Opin Drug Metab Toxicol 2016;12:897–909.

[8] Galeone A, Rotunno C, Guida P, Bisceglie A, et al. Monitoring incomplete heparin reversal and heparin rebound after cardiac surgery. J Cardiothroac Vasc Anesth 2013;27:853–8.

[9] Siddiqui MA, Scott LJ. Recombinant factor VIIa (Eptacog Alfa): a review of its use in congenital or acquired hemophilia and other congenital bleeding disorder. Drugs 2005;65:1161–77.

[10] Logan AC, Yank V, Stafford RS. Off-label use of recombinant factor VIIa in US hospitals: analysis of hospital records. Ann Intern Med 2011;154:516–22.

[11] Leissinger C, Carcao M, Gill JC, Journeycake J, et al. Desmopressin (DDAVP) in the management of patients with congenital bleeding disorders. Hemophilia 2014;20:158–67.

[12] Mannucci P. Desmopressin (DDAVP) in the treatment of bleeding disorders: first 20 years. Blood 1997;20:2515–21.

[13] Fralick M, Kesselheim AS. FDA approval of desmopressin for nocturia. JAMA 2017;317:2059–60.

[14] Aitken SL, Tichy ER. Rh₀(D) immune globulin products for prevention of alloimmunization during pregnancy. Am J Health Syst Pharm 2015;72:267–76.

[15] Ayaches S, Herman JH. Prevention of D sensitization after mismatched transfusion of blood components: towards optimal use of RhIG. Transfusion 2008;48:1990–9.

[16] Hunt BJ. The current place of tranexamic acid in the management of bleeding. Anaesthesia 2015;70(Suppl. 1):50–3.

[17] Cap AP, Baer DG, Orman JA, Aden J, et al. Tranexamic acid for trauma patients: a critical review of the literature. J Trauma 2011;71(Suppl):S9–S14.

Sources of errors in transfusion medicine

INTRODUCTION

All blood products are potentially associated with transfusion risks. These risks are generally categorized as infectious and noninfectious. Infectious risks are minimized by screening of prospective blood donors with a predonation questionnaire and individuals identified at risk are deferred from blood donation. If no infectious disease risk is identified in this screening phase, individuals are accepted for donation. Donated blood is then tested for major infectious disease risk. If testing results are negative, the blood is released for further transfusion. Noninfectious risks of blood transfusions are associated with adverse reactions which take place during or after transfusion. Some of these adverse reactions, for example, acute hemolytic transfusion reaction, bacterial contamination or sepsis, transfusion-related lung injury as well as transfusion-related acute graft-versus-host disease, a delayed adverse event of transfusion, maybe potentially fatal [1,2].

Transfusion of blood products requires pretransfusion testing for selection of the right type of blood for the patient. Routine pretransfusion testing includes determination of the ABO/Rh blood type, detection of antibodies in recipient plasma. This is referred to as type and screen. Prior to transfusion the crossmatch or in vitro compatibility testing is done. A positive antibody screen means that the recipient plasma might react with antigens present on the red cell membrane. Additional testing is performed to identify the antibody in the recipient. The reactivity detected in vitro may be due to multiple factors, starting with sample contamination, clerical or technical errors, or it may result from a true antibody–antigen reactions.

PROPER BLOOD COLLECTION FOR BLOOD BANK TESTINGS

An important source of error is drawing blood from the wrong patient when an individual needs to be tested for a type and screen. Blood sample collection should be performed directly by the phlebotomist or nurse after correct patient identification by full name and hospital identification number (attached wristband). It is very important that the collected blood sample is labeled at the bedside and not elsewhere or prelabeled, due to high potential of mislabeling or sample mix-up. Correct labeling includes patient's full name, hospital number, date and time of collection and phlebotomist's initials. Blood is collected in siliconized plain tubes with no additives or in tubes with potassium ethylenediaminetetraacetic acid (K2-EDTA) as anticoagulant. If the sample is drawn from an IV line, the IV infusion should be stopped 5–10 min prior to blood drawing and the first 10 mL discarded to prevent blood contamination with fluids that may cause hemolysis. A tube should be collected full to ensure an appropriate blood: anticoagulant ratio, inverted 3–10 times for proper mixing and transported immediately to the laboratory. Fewer inversions may lead to clotted samples, whereas excessive shaking, exposure to excessive heat of cold or dry ice, delays in transportation or even routine transportation via pneumatic tube system may all lead to sample rejection. Samples should be collected within less than 72 h from a scheduled transfusion otherwise complement-dependent antibodies may be missed due to complement becoming unstable. New blood samples from recipient are needed for repeat pretransfusion testing every 72 h if new transfusion orders are made. The age of the samples may be extended to up to a month in certain clinical settings, such as preoperative evaluation of elective surgery patients if they have a negative antibody screen and no RBC exposure via pregnancy or blood transfusions within the past 3 months. Blood samples and segments from the transfused units are stored at 1–6°C for 7–10 days posttransfusion [2,3].

Correctly labeled samples accompanied by requisition form are sent to the blood bank and a careful clerical check is performed to confirm that the information on the label and on the transfusion request (requisition form) is identical. Patient's history is reviewed for previous blood transfusion, immunohematology testing or transfusion reactions. Blood samples are rejected based on specific criteria if deemed improperly labeled or inadequate for testing. The presence of hemolysis is perhaps the most common reason for sample rejection in the blood bank since it cannot be distinguished from the antibody-mediated hemolysis, which is considered as positive reaction in most immunohematology assays. A hemolyzed or grossly lipemic sample may cause false-positive results. If both clerical check and sample are acceptable, the specimen is processed by immediate centrifugation and the red blood cells (RBCs) and supernatant are separated. Documentation of all steps from receipt of specimen to testing and result interpretation is performed manually or electronically and records are kept confidential for a period of time in accordance with requirements of federal, state, and accrediting agencies.

SOURCES OF INTERFERENCES IN BLOOD BANK TESTING

Interferences occurring in blood testing can result in potentially fatal errors or cause a delay in supplying blood for transfusion. The sources of interference in testing are generally categorized in preanalytical, analytical, and postanalytical. In general, a large proportion of errors occur in the preanalytical phase of the tests. Preanalytic variables that can influence the test results can be grouped broadly under two categories: (1) related to specimen collection, handling, and processing and (2) related to physiologic factors and patient's endogenous variables, including diseases, circulating antibodies, and drug therapy. Physiologic variables include the effects of age, sex, time, season, and altitude, as well as conditions such as menstruation and pregnancy, and lifestyle. Amongst those, the age, sex, and pregnancy status are very important when making recommendations in transfusion medicine. Other patient-related variables, such as clinical diagnosis, past and current medication, and medical or surgical procedures are important and may be directly related to the testing results; for this reason, knowledge of medical history is required for correctly interpreting test results and issuing the right type of blood. Unlike for other laboratory testing, the fasting or postprandial status, prior exercise, posture during blood drawing, or specific timing of blood sample collection in regard to circadian or physiologic cycles are generally not known to interfere with transfusion medicine assays. Factors related to specimen collection, such as effects of the duration of tourniquet application, the anticoagulant/blood ratio, specimen handling, and processing steps in the preparation of serum/plasma and RBC separation can introduce an important preanalytic variable. If such procedures result in hemolysis or fibrin clots being present in the sample, this may be misinterpreted as a positive reaction.

At the initial time of interpretation, the true cause for test results is not known, thus interferences and errors may look alike. Hemolysis resulted from incorrect drawing appears like hemolysis due to immune causes. If suspected, potential errors, including technical and clerical errors, have to be ruled out for correct test interpretation. Interferences should be distinguished from errors occurring in the process of blood testing and administration. A major potential source of error can occur when the protocols for patient identification or blood drawing are not followed and patient is not properly identified (blood is drawn from a patient different than the one intended), or when the blood sample is collected from the right patient, but mislabeled or blood samples are mixed-up. In these instances, the wrong blood in the tube (WBIT) is tested. Although uncommon, WBIT does not represent a true interference, but an error in the process that requires early recognition and further investigation to prevent such events from reoccurrence. Two blood samples from the same patient are required for ABO/Rh determination prior to transfusion in order to prevent clerical error as well as detecting WBIT. If WBIT is suspected, it should be ruled out by repeating the test on a different sample collected from the same patient.

Methodology used for testing plays a definite role in test interpretation. In general, microcolumn gel technology is known to be more sensitive than tube testing and less sensitive than solid phase adherence technology for detection of clinically significant alloantibodies. The advantage of using a methodology with increased sensitivity is better identification of patterns of reactivity specific to an antibody that might not be elucidated by a less sensitive methodology. Studies showed that some clinically significant alloantibodies identified by solid phase technology displayed a weakly reactive nonspecific pattern in gel testing. Few studies reported, however, that gel technology is less sensitive than the tube test methodology for detecting ABO isohemagglutinins and expected anti-A or -B may be missed. The monoclonal anti-A and -B reagents used in the gel cards are also not known to react with B(A), acquired-B, or polyagglutinable RBCs. On the other hand, reagents and chemicals present in the reaction environment of various testing kits may also result in false-positive reactions and trigger extensive testing to rule out the presence of al-loantibodies. Similarly, patient's autoantibodies can react with reagent RBCs present in the antibody screen and identification panel, and consequently mask detection of reactivity due to an alloantibody. Cold agglutinins and warm autoantibodies (WAA) are well documented for giving a pan-reactivity pattern and masking possible under-lying alloantibodies. Autoantibodies to reagents usually lead to false-negative reac-tions due to reagent consumption or less availability for detection of the true target.

Interferences, defined as alterations in the expected reactivity pattern potentially misleading the interpretation and correct identification of significant findings, can occur in various tests performed in the blood bank, but their correct interpretation requires an integrated review of the case.

INTERFERENCES IN ABO/RH TYPING

ABO discrepancies include (1) disagreement between historical and current blood type, (2) discrepancy between forward and reverse reactions, (3) reactivity weaker than expected (4+) between the RBC antigens and corresponding antibodies, and (4) detection of mixed field (mf) type of reactivity. ABO discrepancies detected in patients to be transfused must be resolved before any blood component is transfused unless blood is urgently needed, in which case group O RBCs or AB plasma are is-sued. Discrepancies occurring in donor samples must be resolved before the blood unit is labeled with a blood type. At the initial read of an ABO discrepancy, it is not known which of the typing reactions, the forward or the reverse, reflects the patient's true ABO type. Knowledge of the patient's age and clinical diagnosis, historical blood type and transfusion history, and the results of other tests are useful hints and required for final interpretation.

Clerical and technical errors must be first ruled out since they are the most com-mon causes for ABO discrepancies. Clerical errors are responsible for more than 95% of fatal transfusion reactions and imply patient or specimen misidentification or blood sample mix-up. Causes of technical errors include failure to follow manu-facturer's instruction, failure to add cells or reagents, incorrect test cell preparation, so that the pipette dispenses a lower amount of A or B cells into the well, improper centrifugation or incubation conditions, and use of contaminated reagents or defec-tive equipment [2,4–6]. Since clerical and technical errors are not uncommon, the best course of action is to repeat testing on a better washed RBC sample and using plasma from the original specimen to see if the discrepancy persists. Repeat blood drawing might be needed, especially if sample mix-up is a concern.

True ABO discrepancies (not due to clerical and technical errors) are generally grouped in four categories:

1. Weak or loss of expected RBC antigen.
2. Presence of unexpected RBC antigen-like reactivity.
3. Weak or loss of expected antibody.
4. Presence of unexpected antibody reactivity.

WEAK OR ABSENT REACTIVITY OF EXPECTED ANTIGEN

Weak A or B antigens are seen in subgroups of A and B blood types, due to age, such as in newborns and the elderly, or during disease process. Certain hematological malignancies (leukemia and Hodgkin's lymphoma) are associated with aberrant transferase formation leading to fewer antigens being formed on the red cell membrane. In contrast, solid organ cancers are associated with an excess of soluble substance similar to A or B antigen which can neutralize the typing reagents and result in a false-negative reaction. Chimerism, the presence of a dual population of cells, is routinely seen post ABO-mismatched stem cell transplant and rarely, due to vascular anastomosis in fraternal twins. In chimerism, ABO/Rh testing is expected to be discrepant within approximately a month posttransplant. Transfusion of non-ABO specific blood, such as massive transfusion of O Rh-negative RBC units in trauma patients, as well as fetal–maternal hemorrhage, also creates a chimeric state. Therefore, resolution of an ABO discrepancy should start with checking the patient's age and clinical condition.

PRESENCE OF UNEXPECTED RBC ANTIGEN-LIKE REACTIVITY

Detection of unexpected antigen may be due to misidentification of rouleaux for agglutination, polyagglutinable RBCs, or interference from other substances causing RBC agglutination, such as Wharton's jelly. Autoagglutination due to cold autoantibodies (CAA) often causes unexpected reactivity (antigen-like in the ABO/RH typing, but also antibody-like in the antibody screen). Antibodies not specific to blood group antigens, but reacting with chemicals or drugs present in the reaction microenvironment may also cause this type of discrepancy. Stem cell transplantation, B(A) phenomenon, and fetal–maternal hemorrhage are other known causes of unexpected antigen reactivity. mf agglutination is usually noted in chimeras indicating a mixed RBC population (such as in posttransfusion, massive transfusion of another blood group, bone marrow transplant). Polyagglutination is a phenomenon suspected when RBCs reacts with most normal adult sera but not with autologous serum. It may be due to exposure of a cryptantigen on the RBC surface due to action of an enzyme associated with a microorganism (acquired microbial polyagglutination), by passive adsorption of microbial structures with antigenic structures similar to A, B, H, T, and Tn antigens, acquired due to a somatic stem cell mutation, or due to inherited conditions.

WEAK OR LOSS OF EXPECTED ANTIBODY

The most frequent cause of this discrepancy is failure to detect anti-B in A or O plasma samples. This can readily be corrected by the traditional tube test, which is usually performed at room temperature (RT) and sometimes at 4°C. Weak or absent antibodies are found in newborns, elderly, immunosuppressed patients, post stem cell transplant, and in severe immunodeficiencies.

PRESENCE OF UNEXPECTED ANTIBODY REACTIVITY

Detection of unexpected reactivity may be due to rouleaux formation, passively acquired antibodies (via transfusions, administration of Rh immunoglobulins, intravenous immunoglobulins [IVIG] or other drugs), and the passenger lymphocytes syndrome. Individuals with an ABO subgroup can develop alloantibodies which react with reagent RBCs of the reverse typing. The classic example is the A2 subgroup with anti-A1 antibody. Positive reaction can also result from cold agglutinins and true alloantibodies reacting with antigens present on the surface RBCs used for testing (such as anti-M, anti-N, anti-P1, and anti-c). A citrate-dependent autoantibody causing errors in blood grouping was also described [7].

INTERFERENCES IN THE DETECTION OF ANTIBODIES

The antibody identification process is not always straightforward. If clerical and technical errors are ruled out, the reactivity observed in vitro might be due to (1) false positives, (2) real antigen–antibody reactions, and (3) nonimmune interactions.

False-positive reactions or pseudoagglutination was described above. Common causes of interferences in this category include the presence of fibrin clots, rouleaux formation, and agglutination due to albumin and plasma expanders. Real antigen–antibody reactions can result from simple or combined presence of allo- or autoantibodies directed either against specific RBC antigens or not related to them.

Autoantibodies typically react with patient's own cells and all reagent RBCs present in the antibody screen and panel in a nonspecific manner. Due to their panreactivity, they may mask the detection of alloantibodies. Distinguishing between autoantibodies versus alloantibodies is essential, but the answer is not always in the positive autocontrol since transfused patients with developing alloantibodies have a positive test. It is rather the reactivity pattern that suggests the type of autoantibody. Autoantibodies reacting only at cold temperatures, including RT, thus in the in vitro testing, but not at 37°C, are known as CAA and generally considered clinically insignificant. They become clinically significant if their reactivity extends beyond 32°C into the body temperature, such as in the case of cold agglutinin disease (CAD), or if the patient undergoes hypothermia, such as in cardiac surgery. The typical CAA is (2+) or (mf) equally pan-reactive with reagent RBCs present in the antibody screen and extended panel. Autoantibodies reacting at 37°C or in the presence of anti-human globulin (AHG phase) are described as WAA. The typical WAA is strong (3+ or 4+) and equally pan-reactive with reagent RBCs present in the antibody screen and extended panel. The clinical significance of WAA is related to their propensity to cause hemolysis. As both CAA and WAA are typically pan-reactive, if present, they may mask or interfere with the detection of alloantibodies. In such cases, further workup is needed to rule out the presence of underlying alloantibodies against major RBC antigens. Passively transfused antibodies originate from plasma products containing alloantibodies (such as IVIG or anti-D) or status post-organ or stem cell transplant (passenger lymphocyte syndrome).

Interferences due to antibodies present in the patient's plasma, not due to RBC antigens were described for a variety of chemicals, antibiotics, potentiators low ionic strength saline (LISS), or other substances (lactose, lactate, melibiose, phenol, sucrose, and thrombin) present in the testing environment or commercial antisera [8–10]. Chemicals that may lead to generation of antibodies reacting with RBCs are Paraben, thimerosal, sugars, ethylenediaminetetraacetic acid (EDTA), inosine, citrate, acriflavine, sodium caprylate, yellow #5, and tartrazine. Antibiotics known to cause production of antibodies that are also capable of reacting with RBCs but not via blood group antigens are penicillin, chloramphenicol, neomycin sulfate, gentamicin, tetracycline, streptomycin, and vancomycin.

At least three mechanisms have been described for the reactivity of RBCs: (1) antigen–antibody complexes may form in the test environment leading to RBC agglutination, (2) antibodies may adsorb onto RBC surface and bind to antigen, and (3) RBC agglutination may be enhanced by the presence of exogenous substances independent of an antigen–antibody reaction. Chemicals and drugs interfere with testing by either covalent (penicillin) or noncovalent link to RBCs. In the latter case, washing may remove the reactivity due to residual plasma or due to antibodies that are noncovalently bound. Antibodies against neomycin, chloramphenicol, gentamicin, hydrocortisone, sugars, dyes (acriflavine, yellow #5 tartrazine), sodium azide, and thimerosal are noncovalently bound, whereas antibodies against penicillin, inosine or EDTA are not removed by washing and additional testing is required. Occasionally, these antibodies may display blood group antigen specificity. Other immune interactions were described for an RBC age-dependent antibody and antibodies to lower oxiranes (ethylene, propylene, and butylene oxides) used in the sterilization of polyvinyl chloride blood donor packs. Sterilization of the outside of the bag with propylene oxide sometimes causes the anticoagulant in the bag to acquire properties that could induce RBCs to acquire a new antigen, termed LOX (lower oxirane). If these antibodies are present, they can create problems in pretransfusion testing and present anomalies in ABO, Rh grouping, and antibody detection.

KEY POINTS

- Routine pretransfusion testing includes determination of the ABO/Rh blood type, detection of antibodies in recipient plasma. This is referred to as type and screen.
- An important source of error is drawing blood from the wrong patient when an individual needs to be tested for a type and screen.
- New blood samples from recipient are needed for repeat pretransfusion testing every 72 h if new transfusion orders are made.
- Blood samples and segments from the transfused units are stored at 1–6°C for 7–10 days posttransfusion.
- The presence of hemolysis is perhaps the most common reason for sample rejection in the Blood Bank since it cannot be distinguished from the antibody-mediated hemolysis, which is considered as positive reaction in most immunohematology assays.

- The sources of interference in testing are generally categorized in preanalytical, analytical, and postanalytical.
- Clerical errors are responsible for more than 95% of fatal transfusion reactions and imply patient or specimen misidentification or blood sample mix-up.
- True ABO discrepancies (not due to clerical and technical errors) are generally grouped in four categories:
  - Weak or loss of expected RBC antigen.
  - Presence of unexpected RBC antigen-like reactivity.
  - Weak or loss of expected antibody.
  - Presence of unexpected antibody reactivity.
- Autoantibodies typically react with patient's own cells and all reagent RBCs present in the antibody screen and panel in a nonspecific manner. Due to their pan-reactivity, they may mask the detection of alloantibodies.
- Autoantibodies reacting only at cold temperatures, including RT, thus in the in vitro testing, but not at 37°C, are known as CAA and generally considered clinically insignificant.
- Autoantibodies reacting at 37°C or in the presence of anti-human globulin (AHG phase) are described as WAA.

REFERENCES

[1] Harmening D. Modern blood banking and transfusion practices. 5th ed. FA Davis Company; 2005.
[2] Roback JD, Combs MR, Grossman BJ, Hillyer CD. Technical manual and standards for blood banks and transfusion services on CD-ROM. 17th ed. American Association of Blood Banks; 2011.
[3] Simon TL, Dzik WH, Snyder EL, Rossi EC, et al. Rossi's principles of transfusion medicine. 3rd ed. Philadelphia: Lippincott Williams & Wilkins; 2002.
[4] Quinley ED. Immunohematology: principles and practice. 3rd ed. Lippincott Williams & Wilkins; 2010.
[5] Issitt PD, Anstee DJ. Applied blood group serology. 4th ed. Durham (NC): Montgomery Scientific; 1998.
[6] Bobryk S, Goossen L. Variation in pipetting may lead to the decreased detection of antibodies in manual gel testing. Clin Lab Sci 2011;24:161–6.
[7] Joshi SR. Citrate-dependent auto-antibody causing error in blood grouping. Vox Sang 1997;72:229–32.
[8] Garratty G. Problems in pre-transfusion tests related to drugs and chemicals. Am J Med Technol 1976;42:209–19.
[9] Garratty G. Screening for RBC antibodies—what should we expect from antibody detection RBCs. Immunohematology 2002;18:71–7.
[10] Garratty G. The significance of complement in immunohematology. Crit Rev Clin Lab Sci 1984;20:25–56.

Index

Printed in the United States
By Bookmasters